T0334399

Process Capability Analysis
Analysis
Estimating Quality

Process Capability Analysis

Analysis

Estimating Quality

Neil W. Polhemus

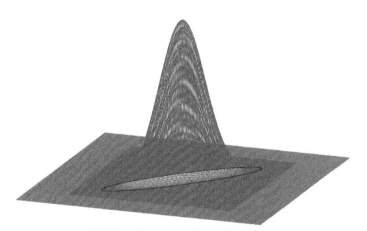

www.statgraphics.com

CRC Press
Taylor & Francis Group
Boca Raton London New York

CRC Press is an imprint of the
Taylor & Francis Group, an **informa** business
A CHAPMAN & HALL BOOK

CRC Press
Taylor & Francis Group
6000 Broken Sound Parkway NW, Suite 300
Boca Raton, FL 33487-2742

First issued in paperback 2020

© 2018 by Taylor & Francis Group, LLC
CRC Press is an imprint of Taylor & Francis Group, an Informa business

No claim to original U.S. Government works

ISBN 13: 978-0-367-57251-8 (pbk)
ISBN 13: 978-1-138-03015-2 (hbk)

Visit the Taylor & Francis Web site at
http://www.taylorandfrancis.com

and the CRC Press Web site at
http://www.crcpress.com

Contents

Preface

Over the last 30 years, I have taught hundreds of courses showing engineers, scientists, and other professionals how to analyze data using Statgraphics and other statistical software programs. Many of these courses covered the fundamentals of statistical process control. While participants were usually familiar with the equations used to compute indices such as C_{pk}, they were often not familiar with how those indices fit into the larger picture of estimating process capability and performance, nor were they always comfortable with how to proceed when assumptions such as normality were not tenable or when multiple variables needed to be analyzed simultaneously.

This book considers the problem of estimating the probability of nonconformities in a process from the ground up. It examines methods based on both attribute data and variable data, considering both classical and Bayesian approaches. For variable data, the book looks at the techniques that were initially developed for data from normal distributions and considers how they must be modified to deal with nonnormal data. The importance of capability indices and their relationship to the percentage of nonconforming items is discussed, as is the use of statistical tolerance limits. Finally, univariate capability analysis is extended to the multivariate situation, which is too often ignored.

I have tried to limit the formulas in this book to those that are necessary to understand the statistical basis for the procedures. Formulas that are only necessary to perform calculations (such as methods for obtaining SPC constants) are not included, since it is assumed that readers will use a statistical software program to do the calculations.

It should also be noted that many statistics in this book are displayed using 6 significant figures. This is not because I believe that so many figures are useful. In fact, I would expect analysts to round off the results in most cases. However, statistics such as the sample mean and standard deviation are often used in subsequent calculations. Carrying too few decimal places into those calculations sometimes has a remarkably large effect on the final results. While this is not an issue for statistical software that carries many significant digits, it can be an issue for readers trying to reproduce the results by hand.

The output presented in this book was produced by Statgraphics Version 18. Appendix B details the steps that are necessary to use that program to generate the output displayed. Other statistical software can be used to perform many of the calculations in this book, although techniques that depend on bootstrapping and Monte Carlo simulation may be difficult to find in other programs.

As you read through this book, you will soon notice that it is not a textbook. Rather, it is a book designed for individuals who have the responsibility of demonstrating that a process is capable of producing goods or services that meet specific requirements. Whether the variable of interest is the diameter of a medical device or the ability of a mass transit system to convey passengers safely from one location to another, the central focus is on applying statistical methods in ways that generate valid estimates of process quality.

Acknowledgments

My interest in statistical methods began while I was a sophomore at Princeton University and was fortunate enough to have as professor Dr. J. Stuart Hunter. He had a way of bringing statistics to life, always beginning his lectures with a story about how he had used what we were about to learn to help improve a real-world process. I later worked with him on various projects, including helping the FAA determine the impact that changing separation between jet routes would have on aircraft collision risk. Stu believed in learning from your data, not trying to make it say what you wanted to hear. I am very grateful to him for all the support he has given me through the years.

I also thank my parents, who sacrificed much so that I could pursue my dreams. I am grateful to Caroline Chopek, who has been instrumental in helping Statgraphics become a widely used tool for quality control and improvement. Thanks also to Seth Wyatt for his hard work in testing the software. I thank my sons Christopher, Gregory, Leland, and Michael for their understanding of the time I needed to complete this project.

Neil W. Polhemus
The Plains, Virginia

Author

Dr. Neil W. Polhemus is chief technology officer for Statgraphics Technologies, Inc., and directs the development of the Statgraphics statistical analysis and data visualization software products. He received his BSE and PhD from the School of Engineering and Applied Science at Princeton University, under the tutelage of Dr. J. Stuart Hunter. Dr. Polhemus spent two years as an assistant professor in the Graduate School of Business Administration at the University of North Carolina at Chapel Hill, where he taught courses on business statistics, forecasting, and quantitative methods. He spent six years as an assistant professor in the Engineering School at Princeton University, where he taught courses on engineering statistics, design of experiments, and stochastic processes. Dr. Polhemus founded Statistical Graphics Corporation in 1980 to develop and promote the Statgraphics software program. In 1983, he founded Strategy Plus, Inc., which developed ExecUStat for managerial statistics. In 1999, the development of Statgraphics was assumed by Statgraphics Technologies, Inc., which also developed StatBeans for statistical analysis in Java and Statgraphics Stratus for statistical analysis in the Cloud.

Chapter 1

Introduction

Process capability analysis refers to a set of statistical methods designed to estimate the capability of a manufacturing or service process to meet a set of requirements or specification limits. The output of the analysis is typically an estimate of the percentage of items or service opportunities that conform to those specifications. If the estimated percentage is large enough, the process is said to be "capable" of producing a satisfactory product or service.

It is customary when studying statistical process control (SPC) to distinguish between two types of data:

1. *Variable data*—measurements made on a continuous scale, such as the dimensions of a manufactured item or the time required to perform a task
2. *Attribute data*—observations made on a nonmeasurable characteristic, usually resulting in a binary decision (good or bad)

This chapter considers methods for summarizing both types of data.

Example 1.1 Medical Devices

Table 1.1 shows the measured diameter of 100 medical devices, randomly sampled from a production process. The diameter of the devices is required to fall within the range 2.0 ± 0.1 mm. Based on this data, we wish to estimate the percentage of items being manufactured by that process that are likely to fall within the required interval.

Example 1.2 Airline Accidents

The U.S. National Highway Safety Administration reported that in 2014, there were 29,989 fatal motor vehicle accidents in the United States. This equates to a fatality rate of 1.07 deaths per 100 million vehicle miles traveled. At the same time, the U.S. Bureau of Transportation Statistics reported the data shown in Table 1.2 for all U.S. air carriers (scheduled and unscheduled) operating under 14 CFR 121. In estimating the quality of service provided by the air carriers, it will be interesting to compare their performance to that of motor vehicles.

The remainder of this chapter examines methods for summarizing data, including both graphical and numerical methods.

1.1 Relative Frequency Histogram

The first step when analyzing any data is to plot it. For variables such as diameter, which are measured on a continuous scale, a relative frequency histogram is very useful. A histogram divides the range of the data into nonoverlapping intervals of equal width and displays bars with height proportional to the number of observations that fall within each interval.

Table 1.1 Measured Diameter of 100 Medical Devices

1.979	1.985	2.021	1.987	1.984	1.985	1.976	1.990	1.999	1.978
1.997	1.964	1.979	1.966	1.979	1.988	1.990	2.002	1.988	1.997
1.978	1.983	1.958	1.968	1.971	1.971	1.956	1.993	2.002	1.983
1.983	1.974	1.990	1.994	1.989	2.053	1.997	1.993	1.988	2.035
2.007	1.985	1.983	1.967	2.005	1.971	1.973	2.002	1.993	2.000
1.987	2.011	1.968	1.977	1.985	2.037	1.980	1.960	1.984	2.004
1.971	2.001	1.965	1.970	1.977	1.984	1.971	2.008	1.996	1.994
1.972	1.989	1.975	1.991	1.976	2.028	1.983	2.045	2.003	1.980
1.972	2.020	2.012	1.983	1.977	1.961	1.969	1.973	1.983	2.008
1.988	1.990	1.996	1.981	1.987	1.972	1.982	1.982	1.986	1.984

Table 1.2 Fatality Statistics for U.S. Air Carriers

Year	Total Accidents	Fatal Accidents	Fatalities Aboard	Flight Hours (Thousands)	Miles Flown (Millions)	Departures (Thousands)
1990	24	6	39	12,150	4,948	8,092
1991	26	4	50	11,781	4,825	7,815
1992	18	4	33	12,360	5,039	7,881
1993	23	1	1	12,706	5,249	8,073
1994	23	4	239	13,124	5,478	8,238
1995	36	3	168	13,505	5,654	8,457
1996	37	5	380	13,746	5,873	8,229
1997	49	4	8	15,838	6,697	10,318
1998	50	1	1	16,817	6,737	10,980
1999	51	2	12	17,555	7,101	11,309
2000	56	3	92	18,299	7,524	11,468
2001	46	6	531	17,814	7,294	10,955
2002	41	0	0	17,290	7,193	10,508

(Continued)

Table 1.2 (Continued) Fatality Statistics for U.S. Air Carriers

Year	Total Accidents	Fatal Accidents	Fatalities Aboard	Flight Hours (Thousands)	Miles Flown (Millions)	Departures (Thousands)
2003	54	2	22	17,468	7,280	10,433
2004	30	2	14	18,883	7,930	11,023
2005	40	3	22	19,390	8,166	11,130
2006	33	2	50	19,263	8,139	10,821
2007	28	1	1	19,637	8,316	10,928
2008	28	2	3	19,127	8,068	10,448
2009	30	2	52	17,627	7,466	9,705
2010	30	1	2	17,751	7,598	9,634
2011	31	0	0	17,963	7,714	9,584
2012	27	0	0	17,722	7,660	9,391
2013	23	2	9	17,693	7,660	9,266
2014	28	0	0	17,599	7,657	9,008

Source: Bureau of Transportation Statistics, Table 2.14: U.S. General Aviation(a) Safety Data, National Transportation Statistics, Department of Transportation, Washington, DC, 2016.

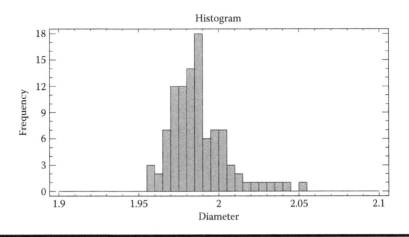

Figure 1.1   Frequency histogram for medical device diameters.

Example 1.1 (Continued)

Figure 1.1 shows a histogram of the medical device diameters. The range covered by the specification limits, 1.9–2.1, has been divided into 40 classes. The bars indicate how many of the 100 sampled devices fall within each class. Notice first that all $n = 100$ devices fall within the specification limits. Second, notice that there are more bars to the right of the peak than there are to the left of the peak, suggesting a lack of symmetry in the distribution of diameter.

While the use of 40 classes for the histogram is somewhat arbitrary, a good rule of thumb is that there should be approximately $10 * \log_{10}(n)$ bars covering the range of the observed values. In this case, $10 * \log_{10}(100) = 20$, which is close to the number of bars displayed in the figure.

1.2 Summary Statistics

Given a sample of n continuous measurements, it is helpful to calculate one or more numerical statistics to summarize the data. A numerical statistic is any number calculated from the data.

Statistics are often used to indicate properties of the data such as central tendency, variability, and shape.

1.2.1 Measures of Central Tendency

The sample of observations will be represented using the notation $\{x_i, i = 1,2,3,\ldots,n\}$. The two most common statistics used to describe the center of the data are the sample *mean* (also called the *average*) and the sample *median*. The sample mean, referred to as $\bar{x}$, is calculated by summing the observations and dividing by n:

$$\bar{x} = \frac{\sum_{i=1}^{n} x_i}{n} \tag{1.1}$$

The median, often referred to as $\tilde{x}$, is calculated by first sorting the observations from smallest to largest. If n is odd, the median is equal to the single observation in the middle. If n is even, the median is the value midway between the middle two observations. If the ith smallest observation is represented by $x_{(i)}$, called the ith *order statistic*, then, if n is odd

$$\tilde{x} = x_{((n+1)/2)} \tag{1.2}$$

If n is even

$$\tilde{x} = \frac{x_{(n/2)} + x_{(1+n/2)}}{2} \tag{1.3}$$

The mean and median quantify the "center" of the data in different ways. While the median is the value that divides the data in half, the mean is equal to the "center of mass". If the observations are plotted along the x-axis, the sample mean is the location where the data values would balance.

Example 1.1 (Continued)

Table 1.3 shows summary statistics for the medical device diameters. There are a total of $n = 100$ observations, resulting in a mean $\bar{x} = 1.98757$ and a median $\tilde{x} = 1.9845$. For data that are positively skewed, it is common for the mean to be somewhat larger than the median since the long right tail of the distribution has a relatively large impact on the calculation of the mean.

Table 1.3 Summary Statistics for 100 Medical Device Diameters

Statistic	
Count	100
Average	1.98757
Median	1.9845
Standard deviation	0.0179749
Coeff. of variation	0.904364%
Minimum	1.956
Maximum	2.053
Range	0.097
Lower quartile	1.976
Upper quartile	1.996
Interquartile range	0.02
Std. skewness	5.03446
Std. kurtosis	4.59981

1.2.2 Measures of Variability

To summarize the magnitude of the variability of the data around its center, three statistics are often calculated: the sample *standard deviation*, the *range*, and the *interquartile range*. The sample standard deviation, referred to as s, is based on the magnitude of the deviations of the observations from the sample mean:

$$s = \sqrt{\frac{\sum_{i=1}^{n}(x_i - \bar{x})^2}{n-1}} \tag{1.4}$$

The greater the variability of the data around the mean, the larger the value of s. If the data come from a normal distribution, $\bar{x}$ and s are sufficient statistics that contain all of the relevant information in the data.

It is also common practice to calculate a *coefficient of variation*. This statistic measures the magnitude of the standard deviation relative to the mean:

$$CV = 100\frac{s}{\bar{x}}\% \tag{1.5}$$

One advantage of the CV is that it has no dimensions, being a percentage ratio of 2 statistics that each have the dimensions of the variable X. The CV is often used when quantifying the amount of error introduced by a measurement process.

Another useful measure of variability is the *range*, calculated by subtracting the minimum value from the maximum value:

$$R = x_{(n)} - x_{(1)} \tag{1.6}$$

The range is sometimes used to estimate the standard deviation of a normal distribution, as will be demonstrated in later chapters. In general, the range is not as good an estimator of spread as the standard deviation, since it emphasizes only the 2 most extreme values. However, for small data sets (no more than 7 or 8 observations), the sample range is nearly as good or "efficient" as the sample standard deviation when estimating the variability of data from a normal distribution.

The *interquartile range* also measures the variability in the data by calculating the distance between the 25th and 75th percentiles. The 25th percentile, also called the *lower quartile* or Q_1, is greater than or equal to 25% of the data values and less than or equal to 75% of the values. The 75th percentile, also called the *upper quartile* or Q_3, is greater than or equal to 75% of the data values and less than or equal to 25% of the values. The interquartile range is

$$IQR = Q_3 - Q_1 \qquad (1.7)$$

The *IQR* can also be used to estimate the standard deviation of a normal distribution.

Example 1.1 (Continued)

As shown in Table 1.3, the medical device data have a sample deviation $s = 0.0179749$, a range $R = 0.097$, and an interquartile range $IQR = 0.02$. The coefficient of variation $CV = 0.904364\%$ shows that the standard deviation is approximately 0.9% of the mean.

1.2.3 Measures of Shape

Two additional statistics are often calculated to measure the shape of the data distribution. The first statistic, called *skewness*, gives an indication of how symmetric the data are. A symmetric distribution has the same shape to the right of its peak as it does to the left. Distributions with longer upper tails than

lower tails are said to be positively skewed, while distributions with longer lower tails are said to be negatively skewed
(Figure 1.2).

The second statistic is called *kurtosis* and measures how
flat or peaked the data distribution is relative to a bell-shaped
normal distribution. Larger values of kurtosis indicate a very
peaked distribution, while smaller values indicate that the distribution is flatter than the normal (Figure 1.3).

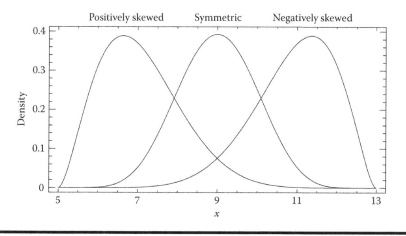

Figure 1.2   Distributions with positive and negative skewness.

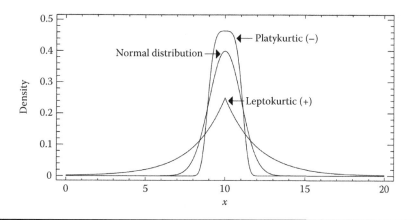

Figure 1.3   Distributions with positive and negative kurtosis.

Widely used statistics for measuring skewness and kurtosis are

$$g_1 = \frac{n \sum_{i=1}^{n} (x_i - \bar{x})^3}{(n-1)(n-2)s^3} \tag{1.8}$$

for skewness, and

$$g_2 = \frac{n(n+1) \sum_{i=1}^{n} (x_i - \bar{x})^4}{(n-1)(n-2)(n-3)s^4} - \frac{3(n-1)^2}{(n-2)(n-3)} \tag{1.9}$$

for kurtosis. Unfortunately, the numerical values of g_1 and g_2 are difficult to interpret. It is usually more helpful to divide each of those statistics by its asymptotic standard error, resulting in *standardized skewness* and *standardized kurtosis* values defined by

$$z_1 = \frac{g_1}{\sqrt{6/n}} \tag{1.10}$$

and

$$z_2 = \frac{g_2}{\sqrt{24/n}} \tag{1.11}$$

In large samples, these statistics will fall within the range −1.96 to 1.96 with 95% probability when the data are random samples from a normal distribution. They may therefore be used as a quick test for normality. Values outside that range are indications that the data probably do not come from a symmetric, bell-shaped, normal distribution.

Example 1.1 (Continued)

The medical device data have a standardized skewness equal to 5.03446, which is well above the expected range for data from a normal distribution. As can be seen from the histogram shown earlier, the distribution has a noticeably longer tail in the positive direction. The standardized kurtosis is also outside the range expected for a normal distribution. Together, these two statistics provide strong evidence that the data are not a random sample from a normal distribution. Chapter 5 describes a formal test for normality called the Shapiro-Wilk test, which should be conducted whenever the standardized skewness and kurtosis are not within the expected range of −1.96 to 1.96.

1.3 Box-and-Whisker Plot

The famous statistician John Tukey developed a very useful graph for variable data called a box-and-whisker plot that displays a 5-number summary of the data. It consists of a box covering the distance between the *lower* and *upper quartiles*, a vertical line at the *median*, and whiskers extending out to the *minimum* and *maximum* values (excluding any unusual points). Any observations that appear to be unusually far removed from the majority of the data, which Tukey called *outside points*, are displayed using separate point symbols, in which case the whiskers extend out to the most extreme points that are not outside points.

Example 1.1 (Continued)

Figure 1.4 shows a box-and-whisker plot for the medical device diameters. Notice that the right whisker extends farther from the box than the left whisker, indicating positive skewness. In addition to the vertical line at the sample median, a small + sign indicates the location of the sample mean. As with most positively skewed distributions, the sample mean is larger than the median. The graph also displays separate point symbols for the five largest observations, which are outside points.

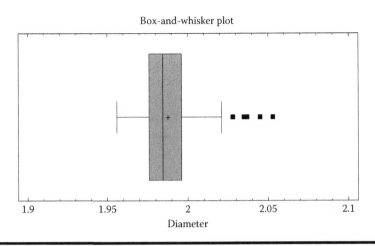

Box-and-whisker plot

Figure 1.4 Box-and-whisker plot for medical device diameters.

Tukey defined two kinds of *outside points*: regular out-side points, which are more than 1.5 times the *IQR* above or below the box, and *far outside points*, which are more than 3 times the *IQR* away from the box. His rule for iden-tifying far outside points is one of the more commonly used tests to determine whether a data sample contains outliers, observations that do not come from the same pop-ulation as the others in the sample. Dawson (2011) showed that, in practice, data sampled from a normal distribution will frequently give rise to ordinary *outside* points (30% of samples from a normal distribution will display at least 1 outside point), but it would be very unusual to see any observations far enough from the central box to be classi-fied as *far outside*, except in samples for which the sample size $n < 10$.

Note: Both types of outside points occur more frequently if the data are skewed. Outside points may therefore indicate either the presence of outliers or the fact that the data come from a nonnormal distribution.

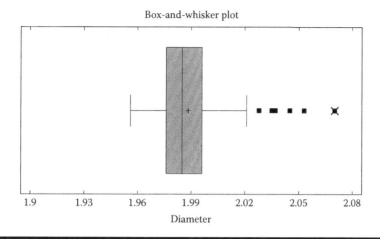

Figure 1.5   Box-and-whisker plot showing far outside point.

Example 1.1 (Continued)

It is common practice to differentiate between ordinary
outside points and *far* outside points. In Figure 1.5, an additional value has been added to the sample with a diameter
of 2.07. It appears as a point symbol with a superimposed
X, indicating that it is a far outside point.

Many analysts like to indicate uncertainty in the location of
the sample mean or median by adding additional features to
the box-and-whisker plot. McGill et al. (1978) suggested cut-
ting a notch in the edge of the box to indicate the width of a
confidence interval for the median. Other authors have suggested using a diamond shape to display a confidence interval
for the median or mean.

Example 1.1 (Continued)

Figure 1.6 shows a modified box-and-whisker plot for the
medical device diameters. The notch in the top and bottom
of the box indicates the width of a 95% confidence interval
for the median diameter.

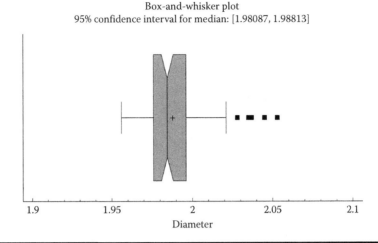

Figure 1.6 Modified box-and-whisker plot showing 95% confidence interval for the median.

1.4 Plotting Attribute Data

When the data are not continuous, different methods need to be employed to summarize it. Consider, for example, the data on air traffic accidents shown in Table 1.2. This data will be used to determine whether the current air transportation system is capable of providing safe travel.

Example 1.2 (Continued)

There are several metrics that might be used to quantify the risk associated with air travel: the total number of accidents, the number of fatal accidents, or the number of fatalities. Furthermore, these quantities could be expressed in terms of miles traveled, hours flown, flight segments, or total trips. A metric often used by the International Civil Aviation Organization (ICAO) is

X = Number of fatal accidents per 100 million flying hours

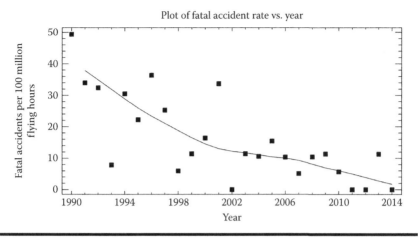

Figure 1.7   Fatal accidents per 100 million flying hours between 1990 and 2014 with robust LOWESS.

Figure 1.7 shows this quantity for each year between 1990 and 2014. Superimposed on the plot is a smoother calculated using the robust LOWESS method developed by Cleveland (1979). LOWESS estimates the smoothed value of Y at any given X by doing a weighted regression of the values closest to X. To make the smoother less sensitive to outliers, a second smoothing is performed after down-weighting values that are far removed from the first smooth. It is clear from the figure that the rate of fatal accidents has been declining steadily over that period.

1.5 Estimating the Percentage of Nonconformities

As mentioned earlier, the primary purpose of performing a capability analysis is to estimate the percentage of items in a population that do not conform to the specifications for a product or service. Those specifications may take the form of an acceptable range, such as 2.0 ± 0.1, a single upper or lower bound, or a more subjective statement about the required attributes for the item.

1.5.1 Proportion Nonconforming

Given a sample of n items from a large population, the critical task is to use those items to estimate the proportion of similar items in the entire population that do not satisfy the product specifications or requirements. Such items are commonly referred to as *nonconforming items*. The proportion of such items will be denoted by

θ = Proportion of nonconforming items in the population.

Several types of estimates are desired:

1. A point estimate $\hat{\theta}$, which gives the best single estimate for that proportion
2. A confidence interval $\left[\hat{\theta}_L, \hat{\theta}_U\right]$, which gives a range of estimates that will contain the true value θ in a stated percentage of similar analyses (often 95%)
3. An upper confidence bound $\hat{\theta}_U$ that does not underestimate the true value of θ in a stated percentage of similar analyses

1.5.2 Defects per Million

When the proportion of nonconforming items is very small, it is useful to express that proportion in terms of the number of items out of every million that do not conform to the specifications. This is commonly referred to as *defects per million* and is related to the proportion of nonconforming items by

$$DPM = 1,000,000\ \theta \qquad (1.12)$$

A related metric for measuring product quality is the *percent yield* given by

$$\%\,yield = 100(1-\theta) \qquad (1.13)$$

The *% yield* is the percentage of items that *do* satisfy the specifications.

In this book, the word "item" will be interpreted broadly. It may represent a physical item such as a medical device, it may represent an encounter with a customer service representative, or it may represent a span of time during which an event such as an aircraft accident could occur. The most important aspect of an "item" is that many exist and each can be classified as either conforming or nonconforming.

1.5.3 Six Sigma and World Class Quality

The acceptable proportion of nonconforming items depends strongly on the product or service being provided, the variable being measured, and the costs associated with nonconformance. Nonconformance of products such as jet engines can be catastrophic. However, under-filling a bottle of soda does not have the same life-and-death consequences. At times, it may be reasonable to accept higher levels of nonconformities for noncritical products if the cost of improving the process exceeds the cost associated with producing a nonconforming item.

A well-known methodology for improving product quality called *Six Sigma* was developed by Motorola in 1986 and has spread over subsequent years to many companies and organizations. As part of that methodology, the originators of Six Sigma extended the notion of "defects per million" to "defects per million opportunities" or *DPMO*. DPMO recognizes that for most products and services, there is more than one opportunity to fail. The formula for DPMO is usually expressed as

$$\text{DPMO} = \frac{1,000,000 \cdot \text{number of defects}}{\text{Number of units} \cdot \text{number of opportunities per unit}}$$

$$(1.14)$$

Table 1.4   Sigma Quality Levels with Associated DPMO and Percent Yield

Sigma Quality Level	DPMO	% Yield
1	691,462	30.9%
2	308,536	69.1%
3	66,807	93.32%
4	6,210	99.38%
5	233	99.977%
6	3.4	99.99966%
7	0.019	99.9999981%

Six Sigma practitioners reserve the term "world class quality" for processes that generate no more than 3.4 DPMO. They also associate a "Sigma Quality Level" with each possible value of DPMO. Processes achieving no more than 3.4 DPMO are said to be operating at the "Six Sigma" quality level, for reasons that will be explained later. Table 1.4 shows various sigma quality levels, their corresponding DPMO, and the corresponding *% yield.*

1.5.4 What's Ahead

Subsequent chapters examine methods for estimating the capability of a process using the following techniques:

- Chapter 2 describes methods for estimating the *proportion* of nonconforming items by directly counting the number of nonconforming items in a sample. This approach is capable of dealing with either *variable* data or *attribute* data.
- Chapter 3 describes methods for estimating the *rate* at which nonconformities are being generated, rather than the proportion of nonconforming items. This applies to

situations in which a single item may have more than one defect or in which unacceptable events occur over a continuous interval.

■ Chapter 4 describes methods for analyzing measurements (*variable* data) that come from a normal distribution. It describes in depth the important concept of *capability indices*.

■ Chapter 5 deals with methods for analyzing measurements that do not come from a normal distribution. It includes three approaches: transforming the measurements so that they do follow a normal distribution, fitting a distribution other than the normal and estimating capability indices based on the fitted distribution, and estimating specially constructed nonnormal capability indices.

■ Chapter 6 describes an alternative approach for dealing with variable data called *statistical tolerance limits*. Statistical tolerance limits bound a specified percentage of a population with a given level of confidence. These limits can be calculated using data from both normal and nonnormal distributions.

■ Chapter 7 describes the concept of *multivariate capability analysis*, where the behavior of more than one variable is considered simultaneously. For processes characterized by multiple variables that are significantly correlated, a multivariate approach will give better estimates of overall process capability than analyzing each variable separately.

■ Chapter 8 considers the important problem of determining how many samples should be obtained in order to provide adequate estimates of process quality. The sample size problem is addressed from the viewpoint of both precision and power.

■ Chapter 9 concludes with a discussion of *control charts* applied to capability analysis. Once a process has been declared to be "capable", these charts monitor continued conformance to the specifications.

References

Bureau of Transportation Statistics. (2016), Table 2.14: U.S. General Aviation(a) Safety Data, National Transportation Statistics, Washington, DC: Department of Transportation.

Cleveland, W.S. (1979), Robust locally weighted regression and smoothing scatterplots, *Journal of the American Statistical Association*, **40**, 829–836.

Dawson, R. (2011), How significant is a boxplot outlier? *Journal of Statistics Education*, **19**, 1–13.

McGill, R., Tukey, J.W., and Larsen, W.A. (1978), Variations of box plots, *The American Statistician*, **32**, 12–16.

Bibliography

ASTM E2281-15. (2015), *Standard Practice for Process Capability and Performance Measurement*, West Conshohocken, PA: ASTM International.

Bothe, D.R. (1997), *Measuring Process Capability: Techniques and Calculations for Quality and Manufacturing Engineers*, New York: McGraw-Hill.

Breyfogle, F.W. III. (2003), *Implementing Six Sigma: Smarter Solutions® Using Statistical Methods*, 2nd edn., New York: John Wiley & Sons.

Carey, R.G. and Lloyd, R.G. (2001), *Measuring Quality Improvement in Healthcare: A Guide to Statistical Process Control Applications*, Milwaukee, WI: ASQ Press.

Chambers, J.M., Cleveland, W.G., Kleiner, B., and Tukey, P.A. (1983), *Graphical Methods for Data Analysis*, New York: Chapman & Hall.

Cleveland, W.G. (1994), *The Elements of Graphing Data*, Monterey, CA: Wadsworth.

Crossley, M.L. (2000), *The Desk Reference of Statistical Quality Methods*, Milwaukee, WI: ASQ Press.

Frigge, M., Hoaglin, D.C., and Iglewicz, B. (1989), Some implementations of the boxplot, *The American Statistician*, **43**, 50–54.

Joglekar, A.M. (2003), *Statistical Methods for Six Sigma in R&D and Manufacturing*, New York: John Wiley & Sons.

Montgomery, D.C. (2013), *Introduction to Statistical Quality Control*, 7th edn., Hoboken, NJ: John Wiley & Sons.

National Traffic Safety Administration. (2016), 2014 motor vehicle crashes: Overview, Traffic Safety Facts Research Note DOT HS 812 246, Washington, DC: U.S. Department of Transportation.

Panda, A., Jurko, J., and Pandova, I. (2016), *Monitoring and Evaluation of Production Processes: An Analysis of the Automotive Industry*, New York: Springer.

Pyzdek, T. (2003), *The Six Sigma Handbook, Revised and Expanded: A Complete Guide for Greenbelts, Blackbelts, & Managers at All Levels*, New York: McGraw-Hill.

Ryan, T.P. (2000), *Statistical Methods for Quality Improvement*, 2nd edn., New York: John Wiley & Sons.

Spiring, F., Leung, B., Cheng, S., and Yeung, A. (2003), A bibliography of process capability papers, *Quality and Reliability Engineering International*, **19**, 445–460.

Tukey, J.W. (1977), *Exploratory Data Analysis*, Boston, MA: Addison-Wesley.

Velleman, P.F. and Hoaglin, D.C. (1981), *Applications, Basics and Computing of Exploratory Data Analysis*, Boston, MA: Duxbury.

Chapter 2

Capability Analysis Based on Proportion of Nonconforming Items

This chapter and the next examine methods for performing a capability analysis using attribute data. Given a sample of *n* items randomly sampled from a large population, each item is inspected and compared to the requirements for the product or service being rendered. Two situations are of interest:

- Situation 1: Each item is classified as either *conforming* or *nonconforming* based on whether or not it meets the specifications for the item.
- Situation 2: The number of nonconformities for each item is noted and the sum of all nonconformities in the sample is calculated. In this situation, more than one nonconformity may be identified on a single item.

This chapter deals with situation 1 and concentrates on estimating the *proportion of nonconforming items*. The next chapter deals with situation 2 and concentrates on estimating the *rate of nonconformities per item*.

2.1 Estimating the Proportion of Nonconforming Items

Suppose a random sample of n items is obtained from a large population. Let

X = Number of nonconforming items in the sample

The probability distribution associated with the random variable X is the binomial distribution, which has a probability mass function equal to

$$p(x) = \binom{n}{x} \theta^x (1 - \theta)^{n-x} \qquad (2.1)$$

where θ is the probability that a randomly selected item does not conform to the specifications. Given that X out of the n items are found to be nonconforming, the maximum likelihood estimate of θ is given by the fraction of nonconforming items in the sample:

$$\hat{\theta} = \frac{X}{n} \qquad (2.2)$$

Equation 2.2 provides the best *point estimate* for θ.

Example 2.1 Estimating the Proportion of Nonconforming Items

Chapter 1 described an example in which $n = 100$ medical devices were sampled from a production process and their diameters were measured. One way to estimate the proportion of nonconforming items is simply to count how many items have diameters outside of the specification limits. In the example, $x = 0$ items were beyond the specification limits, resulting in an estimated proportion of nonconforming items $\hat{\theta} = 0$.

2.1.1 Confidence Intervals and Bounds

While point estimates are useful, they are almost always wrong. Wrong in the sense that rarely, if ever, do they exactly match the quantity that they are trying to estimate. In the medical device example, it should be quite obvious that observing $x = 0$ defects in a sample of $n = 100$ items is hardly sufficient evidence to declare that no nonconforming items are being produced.

It is therefore important that whenever point estimates are provided, their margin of error is also stated. To quantify the margin of error associated with an estimated binomial proportion, a *confidence interval* for θ may be calculated. Confidence intervals are constructed in such a way that they contain the true value of the parameter being estimated a stated percentage of the time. Common practice uses α to represent the proportion of the time that the confidence interval does NOT contain the true value and is usually set to a value such as α = 0.05.

The following formula provides an approximate 100(1 − α)% confidence interval for the proportion of nonconforming items:

$$\left[\frac{v_1 F_{1-\alpha/2,v_1,v_2}}{v_2 + v_1 F_{1-\alpha/2,v_1,v_2}}, \frac{v_3 F_{\alpha/2,v_3,v_4}}{v_4 + v_3 F_{\alpha/2,v_3,v_4}} \right] \tag{2.3}$$

where

$$v_1 = 2X, \quad v_2 = 2(n - X + 1), \quad v_3 = 2(X + 1), \quad v_4 = 2(n - X) \tag{2.4}$$

$F_{p,v,w}$ represents the value of Snedecor's F distribution with v and w degrees of freedom that is exceeded with probability p. The two-sided confidence interval gives both a lower confidence limit (*LCL*) and an upper confidence limit (*UCL*) for θ.

Example 2.1 (Continued)

For the $n = 100$ medical devices where $x = 0$, the 95% confidence interval for θ is

$$\left[0, \frac{2F_{0.025,2,200}}{200 + 2F_{0.025,2,200}}\right] = [0, 0.036217] \qquad (2.5)$$

It may thus be stated with 95% confidence that the average number of medical devices that are not within the specification limits is somewhere between 0 and 36,217 devices per million.

When estimating a quantity such as the proportion of nonconforming items, it may be argued that the lower limit of the confidence interval is not particularly important. The real concern centers on the chance that $\hat{\theta}$ might be underestimating the true proportion, which would lead in practice to more defects than expected. In such cases, a one-sided upper *confidence bound* would be more desirable than a two-sided confidence interval. For example, a one-sided 95% upper confidence bound is the value that underestimates the true proportion only 5% of the time.

A $100(1 - \alpha)$% upper confidence bound for θ is given by

$$\hat{\theta}_U = \frac{v_3 F_{\alpha,v_3,v_4}}{v_4 + v_3 F_{\alpha,v_3,v_4}} \qquad (2.6)$$

Notice that the percentage point of the F distribution changes from $\alpha/2$ to α, resulting in a tighter upper bound than with a two-sided interval. Similarly, a $100(1 - \alpha)$% lower confidence bound for θ is given by

$$\hat{\theta}_L = \frac{v_1 F_{1-\alpha,v_1,v_2}}{v_2 + v_1 F_{1-\alpha,v_1,v_2}} \qquad (2.7)$$

Example 2.1 (Continued)

For the medical device example, the upper 95% confidence bound for θ is

$$\left[\frac{2F_{0.05,2,200}}{200 + 2F_{0.05,2,200}} \right] = 0.029513 \qquad (2.8)$$

It may thus be stated with 95% confidence that the number of medical devices that are not within the specification limit is no more than 29,513 devices per million.

2.1.2 *Plotting the Likelihood Function*

When estimating a binomial proportion, the classical estimate of θ is obtained by maximizing the likelihood function:

$$l(\theta|x) = \binom{n}{x} \theta^x (1-\theta)^{n-x} \qquad (2.9)$$

The likelihood function can be loosely interpreted as the relative chance of obtaining x nonconforming items as a function of the true proportion of nonconforming items. $\hat{\theta}$ corresponds to the value of θ at which the likelihood function attains its maximum value.

Example 2.1 (Continued)

Figure 2.1 shows a plot of the binomial likelihood function for the medical device data ($n = 100$, $x = 0$). The solid curve is the likelihood function. The dotted line corresponds to the location of the maximum likelihood estimate (in this case 0). The dashed line is located at the 95% upper confidence bound (in this case 0.029513). Statistics are displayed at both the maximum likelihood estimate $\hat{\theta}$ and at the upper confidence bound $\hat{\theta}_U$. Estimation of the quality statistics *DPM, Yield, Z, C_{pk}*, and *SQL* is discussed in the following section.

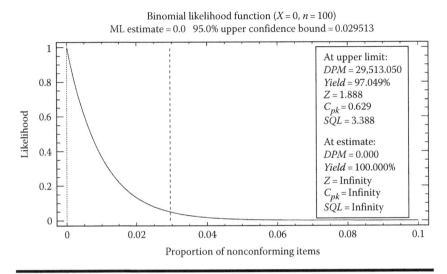

Figure 2.1   Binomial likelihood function for medical device diameters.

2.2 Determining Quality Levels

As discussed in Chapter 1, there are several metrics commonly used to quantify the quality level at which a production process is operating. Equations 1.12 and 1.13 define the *DPM* and *% yield* of a process as direct functions of the proportion of nonconforming items θ. Estimates for those quantities may be obtained by substituting the estimated proportion of nonconforming items into those equations:

$$\text{Estimated } DPM = 1000000\,\hat{\theta} \qquad (2.10)$$

$$\text{Estimated } \% \ yield = 100\,(1 - \hat{\theta})\% \qquad (2.11)$$

An upper bound for *DPM* is obtained by substituting the estimate of the upper bound $\hat{\theta}_U$ into Equation 2.10. A lower bound for *% yield* is obtained by substituting the estimate of the upper bound $\hat{\theta}_U$ into Equation 2.11.

As will be seen in Chapter 4, several other quality indices are often defined when the sample data come from a normal distribution. One is a "*Z*" index, which indicates the value of a standard normal distribution that is exceeded with probability equal to θ. Specifically, if $\Phi(Z)$ is the cumulative standard normal distribution (tabulated in a *Z*-table in most statistics textbooks), then the *Z* index is the value of *Z* that satisfies

$$\Phi(Z) = 1 - \theta \qquad (2.12)$$

For a normal distribution, *Z* quantifies the number of standard deviations between the mean and the nearer specification limit. When considering attribute data as in this chapter, the standard deviation itself is not meaningful. However, Equation 2.12 maintains the relationship between *Z* and θ that exists for variable data and is still useful as an index. General rules of thumb may still be applied, such as those that state that values of $Z > 4$ are desirable.

Another frequently quoted statistic is the capability index C_{pk}, which in the case of a normal distribution divides the distance between the mean and the nearer specification limit by 3σ. C_{pk} is related to the *Z* index by

$$C_{pk} = \frac{Z}{3} \qquad (2.13)$$

Many organizations strive for a $C_{pk} = 1.33$, which for variable data insures that the distance from the mean to the nearer specification limit is at least 4 standard deviations.

Finally, practitioners of Six Sigma define a "Sigma Quality Level", which may be attached to any process. By their definition, a process that achieves an *SQL* of 6 or better is producing product with "world class quality". The SQL can be calculated from *Z* according to

$$SQL = Z + 1.5 \qquad (2.14)$$

The addition of 1.5 to Z comes from the assertion that the mean of most processes is not completely stable but tends to vary around its long-term level by approximately ±1.5 standard deviations.

Table 2.1 shows the relationship between these different quality indices. Although the original meaning of Z and C_{pk} is uninterpretable for attribute data, each may be calculated from $\hat{\theta}$ using Equations 2.12 and 2.13. In such cases, they are called *equivalent* capability indices. Practitioners who are used to interpreting such indices may then use the same rules for both attribute data and variable data and be assured that they correspond to the same level of nonconforming items.

Table 2.1   Relationship between Quality Indices

Z	C_{pk}	SQL	θ	DPM	Yield (%)
0.0	0	1.5	0.500000	500,000	50.0
0.5	0.167	2.0	0.308536	308,536	69.1
1.0	0.333	2.5	0.158655	158,655	84.1
1.5	0.5	3.0	0.066807	66,807	99.32
2.0	0.667	3.5	0.022750	22,750	99.725
2.5	0.833	4.0	0.006210	6,210	99.379
3.0	1.0	4.5	0.001350	1,350	99.865
3.5	1.167	5.0	0.000233	233	99.977
4.0	1.333	5.5	0.000032	31.7	99.9968
4.5	1.5	6.0	0.000003	3.40	99.9997
5.0	1.67	6.5	0.000000	0.29	99.9999

Example 2.1 (Continued)

Figure 2.1 displays the capability indices at both the estimated proportion of defective items $\hat{\theta}$ and at the upper confidence bound $\hat{\theta}_U$. Given the sample data, it has thus been demonstrated with 95% confidence that $DPM \leq 29{,}513$, $Yield \geq 97.05\%$, $Z \geq 1.89$, $C_{pk} \geq 0.63$, and $SQL \geq 3.39$.

In Example 2.1, note the upper confidence bound for θ corresponds to lower confidence bounds for *Yield*, Z, C_{pk}, and *SQL*, since large values of θ correspond to small values of those indices.

2.3 Information in Zero Defects

At first glance, it might seem that there is little information available about the proportion of nonconforming items when a sample of size n results in $x = 0$ nonconformities. In fact, lack of nonconformities does provide useful information about θ. Unfortunately, if the proportion of nonconformities is very small, the sample does not generate a very precise estimate. Of course, the idea that one can obtain a precise estimate of nonconforming proportions on the order of 3.4 defects per million by counting the number of nonconforming items in a sample of size $n = 100$ obviously makes little sense. Even in election polls where the proportion of voters that will eventually choose each candidate is usually in excess of 40%, thousands of potential voters need to be surveyed to get a meaningful result. It is of considerable interest therefore to examine methods for determining what adequate sample sizes are for estimating process capability.

Chapter 8 considers in detail the general problem of determining how large a sample should be taken in order to provide sufficient information about a proportion. Example 2.2 considers the special problem of determining how many samples are enough when the analyst expects to see no nonconformities.

Example 2.2 Sample Size Determination If No Defects Expected

Suppose it is believed that a process for manufacturing medical devices actually produces a very small proportion of nonconforming items. An interesting question to ask is as follows:

> Assuming that no observed items will be beyond the specification limits, how big a sample size n needs to be examined in order to demonstrate with 95% confidence that the proportion of nonconforming items is less than 1 out of 1000?

To solve this problem, the upper confidence bound in Equation 2.6 may be used. A simple algorithm to find n is as follows:

> Step 1: Set $X = 0$ and $n = 1$.
> Step 2: Solve for $\hat{\theta}_U$. If $\hat{\theta}_U \leq 0.001$, stop.
> Step 3: Add 1 to n and repeat Step 2.

Most statistical software can easily solve this problem. As shown in Figure 2.2, the required sample size is $n = 2,993$.

Table 2.2 shows the minimum sample size required such that $x = 0$ nonconformities leads to an upper bound for θ less than various selected values. Proving that the proportion of defective items is small simply by counting the number of nonconforming items can require very large sample sizes.

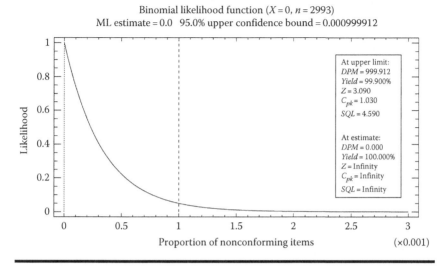

Figure 2.2   Determination of sample size required to achieve upper CL ≤ 0.1% when observing *x* = 0 nonconformities.

Table 2.2   Required Sample Sizes to Achieve Selected Upper Confidence Bounds when Observing 0 Nonconformities in a Sample of Size *n*

Upper Bound	α = 0.10	α = 0.05	α = 0.01
θ = 0.10	n = 22	n = 29	n = 44
θ = 0.05	n = 45	n = 59	n = 90
θ = 0.01	n = 230	n = 299	n = 459
θ = 0.005	n = 460	n = 598	n = 919
θ = 0.001	n = 2,302	n = 2,993	n = 4,601
θ = 0.0005	n = 4,603	n = 5,989	n = 9,206
θ = 0.0001	n = 23,024	n = 29,955	n = 46,048

2.4 Incorporating Prior Information

So far, the methods presented in this chapter have relied solely on the sample of *n* observations to estimate the proportion of nonconforming items. Such an approach assumes that nothing is known about θ before examining the data. It can be argued in many cases that prior information exists about the proportion of nonconforming items. Rarely does one believe a priori that θ is equally likely to be anywhere between 0 and 1. Any prior knowledge about the likely value of θ could be combined with the information obtained from the data to give a potentially more precise estimate.

Bayesian methods let analysts combine prior information with information from the data by quantifying the "degree of belief" concerning possible values of θ. To apply such methods, the analyst begins by expressing his or her knowledge about θ before the data are collected using a prior distribution $p_0(\theta)$. For the binomial distribution, the most common prior distribution is the beta distribution given by

$$p_0(\theta) = \frac{\Gamma(v + w)}{\Gamma(v)\Gamma(w)} \theta^{v-1}(1 - \theta)^{w-1}, \quad 0 \le \theta \le 1 \quad (2.15)$$

The mean of the beta distribution is

$$E(\theta) = \frac{v}{v + w} \quad (2.16)$$

where v and w are parameters that affect the shape of the density function. If $v = 1$ and $w = 1$, the prior distribution is uniform between 0 and 1. If $v > w$, the distribution assigns more probability to large values of θ than to small values. If $v < w$, the distribution assigns more probability to small values of θ than to large values. Figure 2.3 shows beta density functions for various combinations of v and w.

After collecting the data, the prior belief is combined with the likelihood function to obtain a posterior density function $p_1(\theta)$ using

$$p_1(\theta) \propto p_0(\theta)l(\theta|x) \quad (2.17)$$

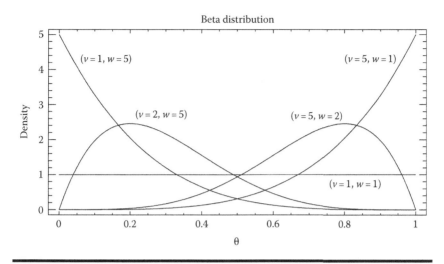

Figure 2.3   Beta density function for different sets of parameters.

If the prior distribution is of the beta form, the posterior density function will also be of the beta form. The two shape parameters of the posterior beta distribution are $v + x$ and $w + n - x$. The mean of the posterior density function provides a Bayes estimator for θ according to

$$\hat{\theta} = \frac{v + x}{v + w + n} \qquad (2.18)$$

2.4.1 *Uniform Prior*

The prior distribution for θ represents the analyst's degree of knowledge or belief about the true proportion of nonconforming items. If one truly knows nothing about θ, it might be reasonable to select a flat or uniform prior where $v = 1$ and $w = 1$. This implies that according to prior knowledge, all values of θ between 0 and 1 are equally likely. After combining the observed data with the prior, this results in a beta posterior distribution with parameters $v = 1 + X$ and $w = 1 + n - X$.

Example 2.3 Using a Uniform Prior

Returning to the medical device diameters, combining a uniform prior distribution with the binomial likelihood function for $x = 0$ and $n = 100$ results in the posterior distribution displayed in Figure 2.4. The mean of the posterior distribution can be used to estimate θ while the upper 95% confidence bound $\hat{\theta}_U$ is found by calculating the 95th percentile of the posterior distribution. Note that the Bayes estimator $\hat{\theta} = 0.009804$ is quite a bit larger than that calculated earlier using the maximum likelihood approach, but the upper 95% confidence bound $\hat{\theta}_U = 0.0292252$ is slightly smaller.

2.4.2 Nonuniform Prior

Using a uniform prior will invariably lead to a larger estimate of the proportion of nonconforming items if θ is small and may thus be unattractive in many capability studies. Where the Bayesian approach is most advantageous is when

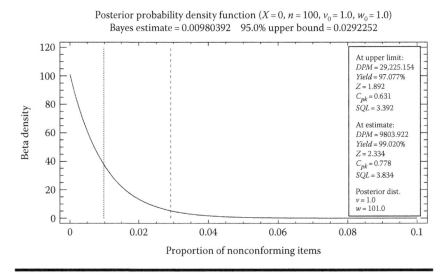

Posterior probability density function ($X = 0$, $n = 100$, $v_0 = 1.0$, $w_0 = 1.0$)
Bayes estimate = 0.00980392   95.0% upper bound = 0.0292252

At upper limit:
$DPM = 29{,}225.154$
$Yield = 97.077\%$
$Z = 1.892$
$C_{pk} = 0.631$
$SQL = 3.392$

At estimate:
$DPM = 9803.922$
$Yield = 99.020\%$
$Z = 2.334$
$C_{pk} = 0.778$
$SQL = 3.834$

Posterior dist.
$v = 1.0$
$w = 101.0$

(y-axis) Beta density
(x-axis) Proportion of nonconforming items

Figure 2.4   Posterior distribution for θ based on uniform prior.

a nonuniform prior is constructed based on the analyst's knowledge of the process. There are two common methods for selecting such a prior distribution:

1. Specifying the mean and standard deviation of the analyst's knowledge about the proportion of non-conforming items and solving for the combination of v and w that yields a beta distribution with those characteristics.
2. Specifying two percentiles of the prior distribution and finding the beta distribution that matches those percentiles.

In the first case, the analyst might indicate that the best guess for θ prior to collecting any data equals 0.01 and that the standard deviation of that prior knowledge is 0.005. In the second case, the analyst might indicate that he or she is 50% certain that $\theta \leq 0.005$ and 90% certain that it is ≤ 0.01. In both cases, the beta shape parameters v and w can be determined to match that prior knowledge.

Example 2.4 Using an Informative Prior

It is a simple matter for statistical software to find a prior distribution that matches the analyst's knowledge. For example, a dialog box similar to that in Figure 2.5 may be used to let the analyst specify either the mean and standard deviation or two percentiles. In Figure 2.5, the analyst has indicated that based on prior knowledge he or she feels that there is a 50% chance that the proportion of nonconforming items is no more than 0.5% and 90% certain that it is no more than 1%. Combining this information with the medical device data results in the posterior distribution shown in Figure 2.6.

After combining the analyst's prior with the data, the posterior distribution has shape parameters $v = 2.93$ and $w = 619.1$. The mean of the posterior distribution provides the Bayes estimate for the proportion of

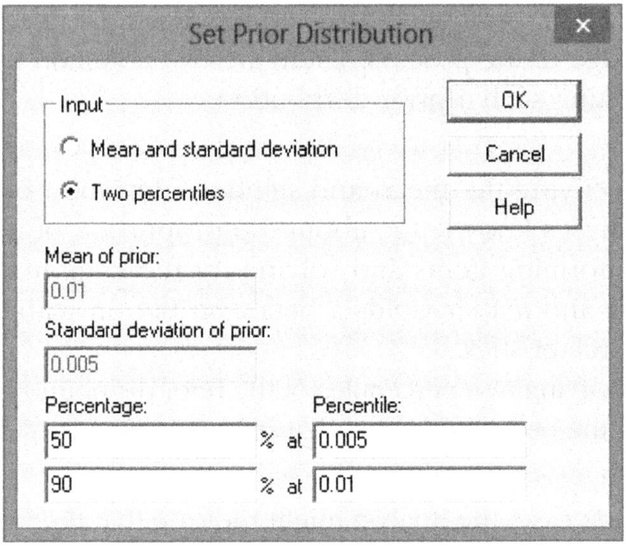

Figure 2.5   Dialog box for determining prior distribution.

nonconforming items, $\hat{\theta} = 0.00471448$. The 95% upper bound is provided by the 95th percentile of the posterior distribution and equals $\hat{\theta}_U = 0.00994061$. Table 2.3 compares these results with those of the maximum likelihood approach. The most noticeable result in that table is the dramatic reduction in the upper bound when specifying an informative prior distribution for θ.

The conclusion reached by the Bayesian analysis depends of course on the choice of the prior distribution. If such analyses are conducted frequently, prior history may give the analyst a good idea of what the likely values of θ are before new data is collected. The data in the current sample may be thought of as providing additional information that enhances the previous knowledge. If used properly, the Bayesian approach can provide more precise estimates than relying solely on the current sample.

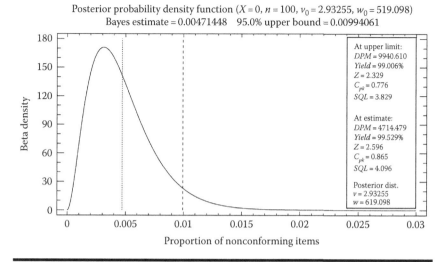

Figure 2.6 **Posterior distribution for proportion of nonconforming medical devices.**

Table 2.3 **Comparison of MLE and Informative Bayesian Approaches**

	MLE Approach	*Bayesian Approach*
Estimated proportion $\hat{\theta}$	0.000000	0.004714
Upper 95% bound for θ	0.029513	0.009941
DPM at upper bound	29,513	9,941

Bibliography

Box, G.E.P. and Tiao, G.C. (1994), *Bayesian Inference in Statistical Analysis*, New York: John Wiley & Sons.

Gelman, A., Carlin, J.B., Stern, H.S., Dunson, D.B., Vehtari, A., and Rubin, D.B. (2013), *Bayesian Statistical Analysis*, 3rd edn., New York: CRC Press.

Guttman, I., Wilks, S.S., and Hunter, J.S. (1982), *Introductory Engineering Statistics*, 3rd edn., New York: John Wiley & Sons.

Johnson, N.L., Kotz, S., and Kemp, A.W. (1993), *Univariate Discrete Distributions*, 2nd edn., New York: John Wiley & Sons.

Chapter 3

Capability Analysis Based on Rate of Nonconformities

This chapter considers the situation in which an analyst obtains a sample of size n and observes X nonconformities. The sample may represent n discrete items, each of which may have multiple nonconformities (such as bubbles in a sheet of glass), or it may represent the size of a continuous sample, such as n days. In either case, the primary parameter of interest is λ, the mean rate of nonconformities per unit.

Example 3.1 Estimating Aircraft Accident Rates

Table 1.2 shows data pertaining to accidents involving U.S. air carriers operating under 14 CFR 121. The U.S. Bureau of Transportation Statistics reported that during the years 2010–2014, such carriers flew 46,882,484 scheduled flight segments lasting 88,727,934 flight hours. Among all of those flights, there were 3 fatal accidents. This chapter considers methods for determining how capable those U.S. carriers are of delivering safe transportation.

3.1 Estimating the Mean Nonconformities per Unit

When a single item can have more than one nonconformity or when the parameter of interest is the occurrence of unacceptable events over a continuous interval, it is necessary to revise the fundamental parameter for measuring quality. In this chapter, the primary parameter of interest will be

$$\lambda = \text{Rate of nonconformities per unit}$$

If a sample of n items is collected and inspected, then the random variable of interest is

$$X = \text{Number of nonconformities in the sample}$$

If a continuous process is observed over a sampling interval of size n, then X represents the number of unacceptable events observed during that interval. Theoretically, X can range from 0 to infinity.

If the nonconformities occur randomly, then the probability distribution associated with the random variable X is the Poisson distribution, which has a probability mass function equal to

$$p(x) = \frac{(\lambda n)^x e^{-\lambda n}}{x!}, \quad x = 0, 1, 2, \ldots \qquad (3.1)$$

If the total number of nonconformities observed in the sample equals X, the maximum likelihood estimate of λ is the number of nonconformities divided by the sample size

$$\hat{\lambda} = \frac{X}{n} \tag{3.2}$$

An approximate $100(1 - \alpha)\%$ confidence interval for λ is given by

$$\left[\frac{\chi^2_{1-\alpha/2,2X}}{2n}, \frac{\chi^2_{\alpha/2,2(X+1)}}{2n} \right] \tag{3.3}$$

where $\chi^2_{p,v}$ is the value of the chi-square distribution with v degrees of freedom that is exceeded with probability p. A $100(1 - \alpha)\%$ upper confidence bound for λ is given by

$$\hat{\lambda}_U = \frac{\chi^2_{\alpha,2(X+1)}}{2n} \tag{3.4}$$

As in Chapter 2, the upper confidence bound is especially useful, since it provides the largest rate of nonconformities per unit that is likely to be true given the observed data.

Example 3.1 (Continued)

To assess the quality of commercial air transportation, there are several measures that could be constructed. On-time performance, baggage handling, and cost are all important. In the end, however, it is the ability of the airlines to get passengers safely from one point to another that really matters.

From 2010 to 2014, U.S. commercial air carriers flew a total of $n = 88{,}727{,}934$ flight hours. There were $x = 3$

fatal accidents. Using Equation 3.2, the estimated rate of fatal accidents per flying hour is

$$\hat{\lambda} = \frac{3}{88,727,934} = 0.0000000338 \qquad (3.5)$$

or slightly less than 3.4 fatal accidents per 100 million flying hours.

Equation 3.4 may be used to calculate an upper confidence bound for the fatal accident rate:

$$\hat{\lambda}_U = \frac{\chi_{0.05,8}^2}{177,455,868} = \frac{15.5073}{177,455,868} = 0.0000000874 \quad (3.6)$$

or approximately 8.74 fatal accidents per 100 million flying hours.

Figure 3.1 shows the likelihood function of the Poisson distribution that describes the observed rate of fatal aircraft accidents, defined by

$$p(\lambda|x) = \frac{(\lambda n)^x e^{-\lambda n}}{x!} \qquad (3.7)$$

The dashed line located at the peak indicates the location of the maximum likelihood estimate, in this case 3.38×10^{-8} or 3.38e^{-8}. The dotted line further to the right corresponds to the upper confidence bound $\hat{\lambda}_U$, in this case 8.74e^{-8}.

3.2 Determining Quality Levels

Measurements of quality based on rates present a unique problem, since the numerical value of the rate is subject to scale. In the aircraft accident example, where the rate measures the occurrence of events per unit time, the choice of units for

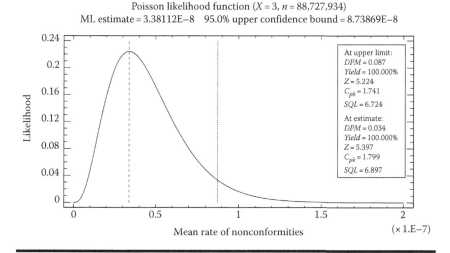

Figure 3.1   Poisson likelihood function for aircraft accidents.

time is arbitrary. Rather than expressing the rate as events per hour, it could just as easily have been expressed as events per minute. The numerical value of λ obviously depends on the unit of time selected. In such cases, it makes little sense to attempt to compute equivalent quality metrics such as Z or C_{pk}, since they would depend on the time unit selected.

In other cases, the unit of measurement is not arbitrary, as when estimating the rate of warranty repairs on dishwashers. In such a case, more than one event might be observed involving the same machine. For such situations, the rate estimate can be converted to meaningful quality indices.

To begin, let θ be the proportion of items in which one or more nonconformities are observed. θ can be calculated from λ by equating it to the probability that $X > 0$ in the Poisson distribution:

$$\theta = 1 - e^{-\lambda} \tag{3.8}$$

or

$$\lambda = -\ln\left(1 - \theta\right) \tag{3.9}$$

If λ is small, θ will be approximately equal to λ since it is very unlikely that two or more events would be observed for the same item.

Example 3.2 Estimating Warranty Repair Rates

Suppose that a study of 1000 dishwashers showed a total of 65 warranty repairs. That results in the estimate $\hat{\lambda} = 0.0650$ repairs per dishwasher. Using Equation 3.4, the 95% upper bound on repairs per dishwasher is $\hat{\lambda}_U = 0.0799$. Substituting both point estimate and upper bound into Equation 3.8 gives $\hat{\theta} = 0.0629$ and $\hat{\theta}_U = 0.0768$, shown as DPM in the statistics box in Figure 3.2.

The estimated values $\hat{\theta}$ and $\hat{\theta}_U$ can then be substituted into the equations presented in Section 2.2 to calculate equivalent values for *DPM, % Yield, Z,* and *SQL*.

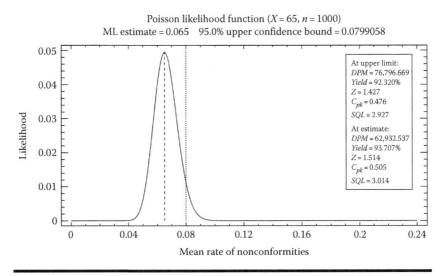

Figure 3.2   Poisson likelihood function for $X = 65$ and $n = 1000$.

Example 3.2 (Continued)

Figure 3.2 shows the Poisson likelihood function for the dishwasher example, the ML estimate, the upper 95% confidence bound for λ, and the calculated quality statistics. Note that the process has a Sigma Quality Level of approximately 3, which most companies would consider unacceptable.

3.3 Sample Size Determination

Chapter 8 discusses in detail the problem of determining an adequate sample size for estimating the rate of nonconformities. As will be seen there, a sample size capable of estimating λ to within any given precision may be found by

1. Picking a level of confidence such as 95%
2. Specifying the expected value of λ
3. Selecting the value of the desired upper bound
4. Using Equation 3.3 or 3.4, solving for n

Again, this is a very simple task for most statistical software.

Example 3.3 Sample Size Determination

Returning to the example of warranty repairs for dishwashers, suppose the records for additional machines will be examined to improve the estimate of the rate of warranty repairs. Assuming that the estimated rate will remain the same, a sample size is desired that will be large enough to reduce the upper bound to under 0.07. Keeping the ratio x/n fixed as close as possible to 0.065, this will require a sample of $n = 8003$ dishwashers as shown in Figure 3.3.

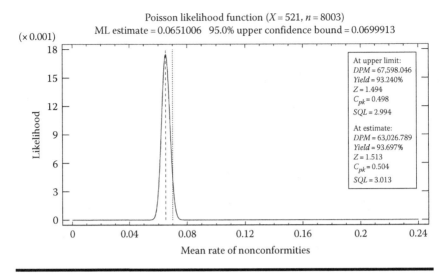

Figure 3.3 Sample size required to achieve 95% upper confidence bound equal to 0.07.

3.4 Incorporating Prior Information

Suppose now that prior knowledge exists about the probable value of λ. Paralleling the development in Chapter 2, this knowledge may be incorporated in the rate estimation process by specifying a prior distribution $p_0(\lambda)$. For the Poisson case, the most commonly used prior distribution is the gamma distribution given by

$$p_0(\lambda) = \frac{\beta^\alpha \lambda^{\alpha-1} e^{-\beta\lambda}}{\Gamma(\alpha)}, \quad 0 \le \lambda \tag{3.10}$$

The mean of the gamma distribution is

$$E(\lambda) = \frac{\alpha}{\beta} \tag{3.11}$$

where α is called the shape parameter and β is called the scale parameter.

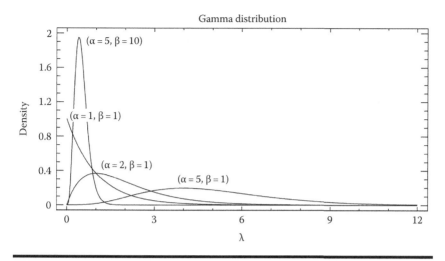

Figure 3.4   Gamma density function for different sets of parameters.

Figure 3.4 shows gamma density functions for various combinations of these parameters.

Selecting a prior distribution of the gamma form and combining it with the data results in a posterior density function $p_1(\lambda)$ which is also of the gamma form. The shape parameter of the posterior gamma distribution is $\alpha + x$ while the scale parameter is $\beta + n$. The mean of the posterior distribution provides a Bayes estimator for λ according to

$$\hat{\lambda} = \frac{\alpha + x}{\beta + n} \tag{3.12}$$

The posterior gamma distribution can be used to provide an upper bound on the true value of λ by calculating the desired percentile.

As with the estimation of proportions in Chapter 2, there are two ways of selecting a prior distribution for λ:

1. Specifying the mean and standard deviation of the analyst's knowledge about the rate of nonconformities and solving for the combination of α and β that yields a gamma distribution with those characteristics

2. Specifying two percentiles of the prior distribution and finding the gamma distribution that matches those percentiles

**Example 3.4 Bayesian Estimation
of Fatal Accident Rate**

Returning to the aircraft accident data, suppose that the analyst is 50% certain that the fatal accident rate per 100 million flying hours is no more than 3 and 90% certain that it is less than 5. Entering that information in the dialog box displayed in Figure 3.5 generates a gamma prior with $\alpha = 5.554$ and $\beta = 1.742e^8$. The posterior probability distribution shown in Figure 3.6 is also a gamma distribution with parameters $\alpha = 8.554$ and $\beta = 2.629e^8$. The Bayes estimate of the accident rate is $\hat{\lambda} = 3.25$ accidents per 100 million flying hours and the upper 95% confidence bound is $\hat{\lambda}_U = 5.27$ accidents per 100 million flying hours. The Bayes estimate is slightly smaller than the estimate obtained using maximum likelihood, while the upper bound is substantially smaller.

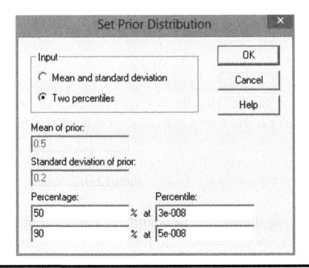

Figure 3.5 Dialog box for determining prior distribution.

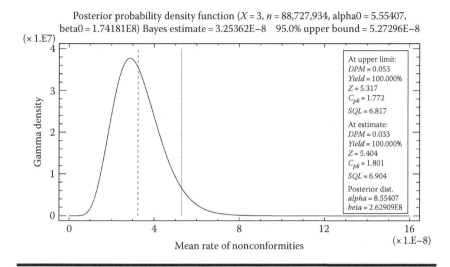

Posterior probability density function ($X = 3$, $n = 88,727,934$, alpha0 = 5.55407, beta0 = 1.74181E8) Bayes estimate = 3.25362E−8   95.0% upper bound = 5.27296E−8

Figure 3.6   Posterior distribution for rate of fatal aircraft accidents.

Bibliography

Box, G.E.P. and Tiao, G.C. (1994), *Bayesian Inference in Statistical Analysis*, New York: John Wiley & Sons.

Cox, D.R. and Lewis, P.A.W. (1966), *The Statistical Analysis of Series of Events*, New York: Chapman & Hall.

Gelman, A., Carlin, J.B., Stern, H.S., Dunson, D.B., Vehtari, A., and Rubin, D.B. (2013), *Bayesian Statistical Analysis*, 3rd edn., New York: CRC Press.

Guttman, I., Wilks, S.S., and Hunter, J.S. (1982), *Introductory Engineering Statistics*, 3rd edn., New York: John Wiley & Sons.

Johnson, N.L., Kotz, S., and Kemp, A.W. (1993), *Univariate Discrete Distributions*: 2nd edn., New York: John Wiley & Sons.

Chapter 4

Capability Analysis of Normally Distributed Data

Earlier chapters concentrated on estimating the capability of a process by inspecting samples from that process and counting the number of nonconformities. This direct estimation of θ, the proportion of nonconforming items, requires relatively few assumptions about the distribution of any variables that might be measured to help determine conformity. However, it throws away important information about how close or far the samples are from the specification limits. Hence the sample sizes required to get precise estimates of θ using such a direct approach are usually quite large.

If the specifications for a product or service are based on a continuous variable X, such as the diameter of a medical device, precise estimates may be obtained from much smaller samples by first modeling the probability distribution of X. Often, it is reasonable to assume that X follows a *normal* or *Gaussian* distribution. This chapter discusses methods for estimating process capability when the assumption of normality holds. Chapter 5 discusses modifications that must be made when the data cannot be assumed to come from a normal distribution.

4.1 Normal Distribution

A common model for the probability distribution of a continuous random variable X is that it follows a normal distribution, defined by a probability density function $f(x)$ that resembles a bell-shaped curve:

$$f(x) = \frac{1}{\sqrt{2\pi\sigma}}\, e^{-\frac{(x-\mu)^2}{2\sigma^2}} \tag{4.1}$$

The normal distribution is defined by two parameters: the mean μ, which indicates the center of mass of the distribution, and the standard deviation σ, which indicates the spread or dispersion. Figure 4.1 shows the normal density function for several values of μ and σ.

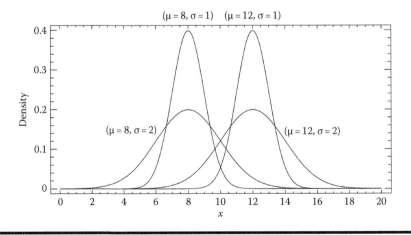

Figure 4.1 Normal distribution with different means and standard deviations.

4.2 Parameter Estimation

Given a random sample of n measurements taken from a normal distribution, the maximum likelihood estimate of the mean is given by $\bar{x}$, the sample mean:

$$\hat{\mu} = \bar{x} = \frac{\sum_{i=1}^{n} x_i}{n} \qquad (4.2)$$

The most common estimate of the standard deviation is given by s, the sample standard deviation, which is calculated from the deviations of the observations around the sample mean:

$$\hat{\sigma} = s = \sqrt{\frac{\sum_{i=1}^{n} (x_i - \bar{x})^2}{n-1}} \qquad (4.3)$$

Example 4.1 Fitting a Normal Distribution

The medical device data introduced in Chapter 1 consist of the measured diameter of $n = 100$ items. Figure 4.2 shows a frequency histogram of the data together with the normal density function with mean $\mu = 1.98757$ and standard deviation $\sigma = 0.0179749$. These parameters match the sample estimates calculated from the data. This normal distribution will be referred to as the *fitted* distribution. Tall vertical lines are drawn at the target value $T = 2.0$ and at the specification limits. Shorter vertical lines are plotted at the mean plus and minus 3 standard deviations, which covers 99.73% of the area under the normal density function.

Confidence intervals can also be obtained for the mean and standard deviation. Confidence intervals quantify the sampling

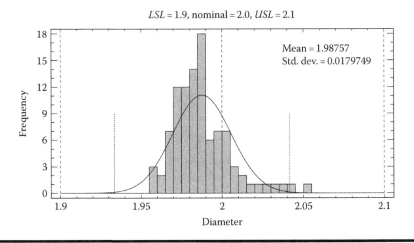

Figure 4.2   Normal distribution fit to medical device diameters.

error associated with the estimates of μ and σ. Any values within the confidence intervals may be considered plausible values for the parameters of the population from which the data were sampled.

A 100(1 − α)% confidence interval for the population mean is given by

$$\bar{x} \pm t_{\alpha/2,n-1} \frac{s}{\sqrt{n}} \tag{4.4}$$

where $t_{\alpha/2,n-1}$ represents the value of Student's t distribution with $n − 1$ degrees of freedom that is exceeded with probability $\alpha/2$. A 100(1 − α)% confidence interval for the standard deviation is given by

$$\left[\sqrt{\frac{(n-1)s^2}{\chi^2_{\alpha/2,n-1}}}, \sqrt{\frac{(n-1)s^2}{\chi^2_{1-\alpha/2,n-1}}} \right] \tag{4.5}$$

where $\chi^2_{\alpha/2,n-1}$ represents the value of the chi-square distribution with $n − 1$ degrees of freedom that is exceeded with probability $\alpha/2$.

Table 4.1   Confidence Intervals for Mean and Standard Deviation of Medical Device Diameters

Confidence Intervals for Diameter
95.0% confidence interval for mean: 1.98757 ± 0.00356661 [1.984, 1.99114]
95.0% confidence interval for standard deviation: [0.0157821, 0.020881]

Example 4.1 (Continued)

Table 4.1 shows 95% confidence intervals for the mean and standard deviation of the medical device diameters. With 95% confidence, it may be stated that the true mean μ is between 1.984 and 1.991 while the standard deviation σ is between 0.0158 and 0.0209.

4.3 Individuals versus Subgroup Data

When performing a process capability analysis, data are usually collected in one of two ways:

1. *Individuals data*: Items are selected and examined one at a time, where each item is assumed to be independent of the others. For example, imagine a manufacturing process in which a single item is randomly selected from each shift's production. For a continuous process, samples might be taken at periodic intervals such as once an hour.
2. *Subgroup data*: When items are naturally segmented into batches or lots, a selected number of items might be randomly selected from each segment. Samples taken from the same segment or at the same time are commonly referred to as belonging to the same *subgroup*.

For example, if the medical devices are produced in lots, 5 devices might be obtained from each lot. When creating subgroups, it is important that the items be selected randomly so that they are representative of the entire segment.

Example 4.2 Analysis of Subgroup Data

Suppose the 100 medical devices described in Chapter 1 consist of 5 items randomly sampled from each of 20 lots. It would then be useful to plot the data using a *Tolerance Chart* as shown in Figure 4.3. A tolerance chart plots the data in each subgroup at a separate location along the X-axis, displaying the range of the data within each subgroup using a vertical line. Horizontal lines are added at the target value and the upper and lower specification limits.

If the subgroup sizes are large or the data are heavily rounded, many of the individual points may be hidden by other points. One solution to such an overplotting problem is to jitter the points, which adds a small amount of random offset to each point in the horizontal direction. Figure 4.4 shows the subgroup data with jittering.

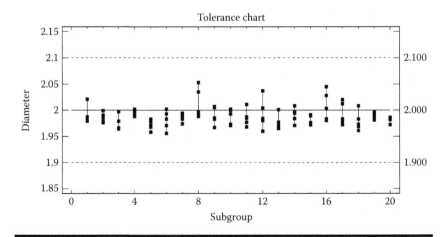

Figure 4.3 Plot of medical device diameters by subgroup with target and specification limits.

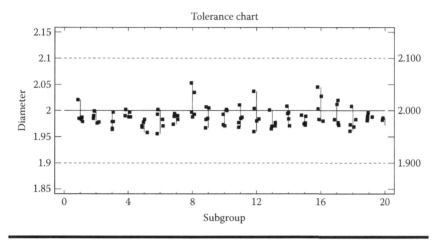

Figure 4.4   Plot of medical device diameters by subgroup with horizontal jittering.

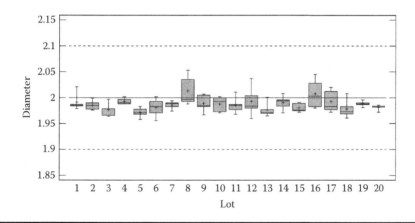

Figure 4.5   Plot of medical device diameters with box-and-whisker plots for each subgroup.

Alternatively, box-and-whisker plots could be displayed for each subgroup as shown in Figure 4.5. Such a plot is very good at visualizing any changes in the level or within-group variability of the data over time.

4.3.1 Levels of Variability

It is common practice when estimating the ability of a process to meet specification limits to estimate two levels of variability: *short-term variability* and *long-term variability*. For subgroup data, short-term variability refers to differences among items within the same subgroup. For individuals data, short-term variability refers to differences between observations collected close together in time. For both types of data, long-term variability refers to differences among items produced during the entire period over which data are collected.

Consider for a moment the case where data are selected in subgroups. If the production process is in a perfect state of statistical control, then all segment means will be identical and the variability within each segment will be the same. In such cases, long-term and short-term variability will be identical. However, if the process mean is not perfectly constant but fluctuates over time, then the segment means will not all be the same. Consequently, the variability of items within the same segment will be less than that of the process as a whole. For individuals data, the effect of a fluctuating process mean causes items produced close together in time to show less variability than those produced over a longer time frame.

To be specific, let $X_{t,j}$ be the observation obtained from the *j*th item sampled at time period *t*. The index *t* refers to the subgroup number. A useful statistical model for the data is

$$X_{t,j} = \mu + \varepsilon_t + \varepsilon_{t,j} \tag{4.6}$$

where
    μ is the process mean
    ε_t is a random between-subgroup error or deviation that has a standard deviation equal to $\sigma_{between}$
    $\varepsilon_{t,j}$ is a random within-subgroup error that has a standard deviation equal to σ_{within}

Assuming that the between-subgroup and within-subgroup errors are independent, the overall standard deviation of the process is then

$$\sigma_{overall} = \sqrt{\sigma_{between}^2 + \sigma_{within}^2} \tag{4.7}$$

4.3.2 Capability versus Performance

If a process is in a strict state of statistical control, the variability among items from the same subgroup will be the same as the variability among items from different subgroups. In such cases, $\sigma_{between} = 0$. Since strict statistical control is rare, statistics calculated from estimates of σ_{within} are said to express the *capability* of the process rather than what is actually being achieved. In contrast, statistics calculated from estimates of $\sigma_{overall}$ are said to express the *performance* of the process.

4.3.3 Estimating Long-Term Variability

The long-term or overall variability in the process is usually estimated by calculating the sample standard deviation of all observations in the data set. If s is the sample standard deviation of the n observations, then

$$\hat{\sigma}_{long\text{-}term} = s \tag{4.8}$$

Occasionally, especially when the total number of observations n is small, a bias correction is applied to s, since the expected value of s is less than the true value of σ. The bias-corrected estimate is given by

$$\hat{\sigma}_{long\text{-}term} = \frac{s}{c_4(n)} \tag{4.9}$$

where $c_4(n)$ is a constant that is tabulated in most books on statistical process control, such as Montgomery (2013). Since $c_4(n)$ is less than 1, the bias correction results in a larger estimate.

4.3.4 Estimating Short-Term Variability from Subgroup Data

Assume that observations have been collected from m subgroups. Let n_j be the number of observations in the jth subgroup. Let $\bar{x}_j$ be the sample mean of the observations within subgroup j. Let s_j be the sample standard deviation of the observations in that subgroup and let R_j be their range. The overall process mean is estimated by calculating a weighted average of the subgroup means:

$$\hat{\mu} = \bar{x} = \frac{\sum_{j=1}^{m} n_j \bar{x}_j}{\sum_{j=1}^{m} n_j} \tag{4.10}$$

The short-term variability (the variability within the subgroups) can be estimated in several ways. Five estimates are used in common practice:

1. An estimate based on the pooled within-group standard deviation
2. An estimate based on the weighted average of the subgroup ranges
3. An estimate based on the weighted average of the subgroup standard deviations
4. A bias-corrected version of estimate 1
5. A bias-corrected version of estimate 3

Approach 1 is based on a standard one-way analysis of variance. It pools the standard deviations within each of the subgroups to estimate the within-subgroup standard deviation:

$$\hat{\sigma}_{short\text{-}term} = \sqrt{\frac{\sum_{j=1}^{m}(n_j - 1)s_j^2}{\sum_{j=1}^{m}(n_j - 1)}} \tag{4.11}$$

If the data within each subgroup come from a normal distribution, then the above estimate follows a chi-square distribution with degrees of freedom $\upsilon = \sum_{j=1}^{m}(n_j - 1)$. A bias correction can be applied if desired:

$$\hat{\sigma}_{short\text{-}term} = \frac{\sqrt{\frac{\sum_{j=1}^{m}(n_j - 1)s_j^2}{\sum_{j=1}^{m}(n_j - 1)}}}{c_4(1+v)} \tag{4.12}$$

Despite the attractive statistical properties of the above estimates, other estimators are more commonly used. Since much of the methodology for calculating process capability was developed prior to the widespread availability of digital computers, alternative methods were sought that reduced the amount of calculation required. In addition, since process capability studies are often done in conjunction with statistical process control charts, estimates were developed that used the calculated values needed to create those charts.

Since the easiest subgroup control charts to construct by hand are X-bar and R charts, which plot the subgroup means and ranges, a widely used estimate of the short-term sigma is based on the average subgroup range. Given m subgroups, the estimated short-term sigma is

$$\hat{\sigma}_{short\text{-}term} = \frac{\sum_{j=1}^{m} f_j^2 R_j / d_2(n_j)}{\sum_{j=1}^{m} f_j^2} \tag{4.13}$$

where

$$f_j = \frac{d_2(n_j)}{d_3(n_j)} \tag{4.14}$$

and d_2 and d_3 are SPC constants similar to c_4.

Another potential estimate of the short-term sigma is based on the average of the subgroup standard deviations. Given m subgroups, the short-term sigma may be estimated using

$$\hat{\sigma}_{short\text{-}term} = \frac{\sum_{j=1}^{m} n_j s_j}{\sum_{j=1}^{m} n_j} \tag{4.15}$$

A bias correction may also be applied to this estimate as follows:

$$\hat{\sigma}_{short\text{-}term} = \frac{\sum_{j=1}^{m} \left(\frac{c_4(n_j)}{1 - c_4(n_j)^2} \right) s_j}{\sum_{j=1}^{m} \left(\frac{c_4(n_j)^2}{1 - c_4(n_j)^2} \right)} \tag{4.16}$$

Example 4.3 Estimating Short-Term and Long-Term Variability from Subgroups

Table 4.2 shows the different estimates obtained if the medical device data are divided into m = 20 subgroups, each containing n_j = 5 consecutive observations. There are two conclusions that can be made from the estimates displayed in that table:

1. Both long-term estimates are greater than all of the short-term estimates, indicating that there might an additional contribution to the process variance caused by variability between the subgroup means. In fact, a one-way ANOVA performed on the data shows significant differences between the subgroup means at the 5% significance level.
2. The short-term estimates based on the pooled standard deviation are larger than those based on weighted averages of the subgroup ranges or standard deviations. This is not unexpected, since Equation 4.11 is based on a weighted average of the squared standard deviations.

Table 4.2   Estimates of Long-Term and Short-Term Variability Using Subgrouped Medical Device Diameters

Method	Equation	Estimated Sigma
Long-term	(4.8)	0.0179749
Long-term (bias corrected)	(4.9)	0.0180203
Short-term, pooled std. deviation	(4.11)	0.0168006
Short-term, pooled std. deviation (bias corrected)	(4.12)	0.0168532
Short-term, average subgroup range	(4.13)	0.0159071
Short-term, average subgroup std. deviation	(4.15)	0.0151362
Short-term, average subgroup std. deviation (bias corrected)	(4.16)	0.0161027

Most statistical software lets you select the estimation methods that you want to use. Note that it is very important to establish a protocol that specifies which methods will be used for estimating the long-term and short-term standard deviations and stick with them, regardless of whether the conclusions work out as desired. As John Tukey cautioned, one should avoid *data snooping*, which is deciding which method to apply after having already looked at the results.

4.3.5 Estimating Short-Term Variability from Individuals Data

If the data to be analyzed are obtained one at a time rather than in subgroups, other methods for estimating short-term variability must be applied. Three methods are in common practice: estimating sigma based on the *average moving range*, estimating sigma based on the *median moving range*, and estimating sigma using the *mean squared successive difference* (MSSD). Each method involves looking at successive data points X_{t-1} and X_t and first calculating the differences between them:

$$D_t = X_t - X_{t-1}, \quad t = 2, 3, \ldots, n \qquad (4.17)$$

Although the observations will not have been obtained at exactly the same time, or necessarily from the same batch or lot, they will be close enough together in time that the differences D_t will exhibit primarily short-term variability.

The absolute values $|D_t|$ are often referred to as the moving range of 2 or MR(2), since they equal the range of each pair of successive observations. A popular estimate of the short-term variability based on the moving range is

$$\hat{\sigma}_{short-term} = \frac{\sum_{t=2}^{n} |D_t| / (n-1)}{1.128} \qquad (4.18)$$

The numerator of (4.18) is the average moving range. The constant in the denominator, which converts the moving range to an estimate of sigma, equals $d_2(2)$ since the estimate is based on the average range of groups of 2 consecutive observations.

The short-term standard deviation can also be estimated from the median of the moving ranges according to

$$\hat{\sigma}_{short\text{-}term} = \frac{Median\left(D_t\right)}{0.954} \qquad (4.19)$$

The constant in the denominator is referred to as $d_4(2)$.

Finally, the short-term sigma can be estimated from the mean of the squared successive differences using

$$\hat{\sigma}_{short\text{-}term} = \sqrt{\frac{\sum_{t=2}^{n} D_t^2 / \left(n-1\right)}{2}} \qquad (4.20)$$

A bias correction may also be applied to the above estimate:

$$\hat{\sigma}_{short\text{-}term} = \sqrt{\frac{\sum_{t=2}^{n} D_t^2 / \left(n-1\right)}{2}} \div c_4'\left(n\right) \qquad (4.21)$$

using an additional SPC constant known as c_4 prime.

Example 4.4 Estimating Short-Term Variability from Individuals Data

The absolute values of the successive differences are often plotted on a moving range chart. Figure 4.6 displays an MR(2) chart for the medical device diameters. The centerline of the chart is the average moving range, while the upper limit is located at the average plus 3 times the estimated standard error.

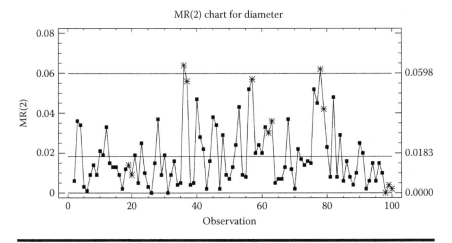

Figure 4.6 Moving range chart for medical device diameters.

Table 4.3 Estimates of Short-Term Variability Using Individual Medical Device Diameters

Method	Equation	Estimated Sigma
Short-term, average moving range	(4.18)	0.0162350
Short-term, median moving range	(4.19)	0.0146751
Short-term, squared successive differences (SSD)	(4.20)	0.0169415
Short-term, SSD (bias corrected)	(4.21)	0.0170071

Table 4.3 shows the different estimates of the short-term variability that would be obtained for the medical device data using the different estimation methods. The estimates based on the mean squared successive difference are the largest, while that based on the median moving range is the smallest.

4.4 Estimating the Percentage of Nonconforming Items

Under the assumption that the observations are random samples from a fitted normal distribution, the mean and standard deviation are sufficient statistics for representing their probability distribution. Given estimates of these two parameters, it is possible to estimate the proportion of nonconforming items θ by calculating the probability that a random variable that follows such a distribution will be outside of the specification limits. This involves using $\hat{F}(x)$, the fitted cumulative normal distribution, which estimates the probability of selecting an item that has a diameter less than or equal to x.

The proportion of nonconforming items is estimated by summing the estimated proportion of items below the lower specification limit and the estimated proportion above the upper specification limit:

$$\hat{\theta} = \hat{F}(LSL) + \left[1 - \hat{F}(USL)\right] \qquad (4.22)$$

If there is no lower specification limit, then $\hat{F}(LSL) = 0$. If there is no upper specification limit, then $\hat{F}(USL) = 1$.

Example 4.5 Capability Analysis of Medical Device Diameters

Table 4.4 shows the output of a capability analysis applied to the medical device diameters, treating the $n = 100$ observations as individuals. The column labeled *Estimated Beyond Spec* gives the estimated tail areas as percentages. The overall estimated proportion beyond the specification limits is $\hat{\theta} = 0.00000055$. If in fact the medical device diameters come from a normal distribution, this provides a best estimate for the proportion of devices being produced that will not be within the specification limits.

Table 4.4 bases its calculations on the estimated long-term standard deviation calculated using Equation 4.8, which uses the sample standard deviation of all n observations. It thus estimates the proportion of nonconforming items produced over a period comparable to that in which the sample was taken. If this period is representative of the entire production, then it also estimates overall performance.

Table 4.4   Capability Analysis of Individual Medical Device Diameters

Data variable: Diameter				
Transformation: None				
Distribution: Normal	Sample size = 100			
	Mean = 1.98757			
	Std. dev. = 0.0179749			
6.0 sigma limits	+3.0 sigma = 2.04149			
	−3.0 sigma = 1.93365			
Specifications	*Observed Beyond Spec*	*Z-Score*	*Estimated Beyond Spec*	*Defects Per Million*
USL = 2.1	0.000000%	6.25	0.000000%	0.00
Nominal = 2.0		0.69		
LSL = 1.9	0.000000%	−4.87	0.000055%	0.55
Total	0.000000%		0.000055%	0.55

4.5 Estimating Quality Indices

Chapter 1 introduced two quality indices that can be calculated directly from θ. The first index, *DPM*, expresses the proportion of nonconforming items as defects per million:

$$DPM = 1,000,000 * \theta \qquad (4.23)$$

The second, *percent yield*, is the percentage of items that fall within the specification limits:

$$\% \, yield = 100(1 - \theta) \qquad (4.24)$$

Many other capability indices have been developed over the years, as described in this section.

4.5.1 Z Indices

Many of the most popular capability indices are based on the mean and standard deviation of the normal distribution. One such index is Z. Z measures how close the process mean is to the specification limits in multiples of the process standard deviation. If a lower spec exists, then the Z index for the lower specification limit is defined by

$$Z_{lower} = \frac{\mu - LSL}{\sigma} \tag{4.25}$$

If an upper spec exists, then the Z index for the upper specification limit is defined by

$$Z_{upper} = \frac{USL - \mu}{\sigma} \tag{4.26}$$

The distance to the nearer specification limit is the smaller of the two one-sided Z indices:

$$Z_{min} = \min\left(Z_{lower}, Z_{upper}\right) \tag{4.27}$$

A value of $Z_{min} = 4$ would indicate that the nearer specification limit is 4 standard deviations from the process mean.

If the values of Z are known, the probability of being beyond the specification limits may be calculated using the standard normal distribution. In particular, if $\Phi(z)$ is the cumulative standard normal distribution, then the proportion of nonconforming items θ is given by

$$\theta = \Phi\left(-Z_{lower}\right) + \left[1 - \Phi\left(Z_{upper}\right)\right] \qquad (4.28)$$

Example 4.5 (Continued)

Table 4.4 contains a column labeled *Z-Score*. This column measures the distance between a particular location (such as the USL, target value, or LSL) and the process mean by subtracting the mean from that location and dividing the result by the standard deviation. The results in that table indicate that the USL is 6.25 standard deviations above the estimated process mean, the target value is 0.69 standard deviations above the estimated process mean, and the LSL is 4.87 standard deviations below the estimated process mean. Note that the *Z*-score for the USL equals Z_{upper}, while the *Z*-score for the LSL equals $-Z_{lower}$.

Both the lower and upper *Z*-scores are in excess of 4.8 in absolute value, meaning that the distance between the sample mean $\bar{x}$ and the specification limits is at least 4.8 times the sample standard deviation s. This corresponds to less than 1 item per million (DPM) outside of the specification limits.

The *Z*-scores displayed in Table 4.4 are based on an estimate of long-term variability. If a short-term estimate of sigma is substituted into Equations 4.25 and 4.26, different results will be obtained. Whereas the long-term sigma quantifies *performance* over the entire sampling period, the short-term sigma quantifies the *capability* of producing a quality product once any long-term changes in the process parameters have been eliminated.

Table 4.5 Comparison of Capability and Performance for Medical Device Diameters

	Short-Term Capability	*Long-Term Performance*
Sigma	0.016235	0.0179749
Z_{upper}	6.92514	6.25484
Z_{lower}	5.39389	4.8718
Z_{min}	5.39389	4.8718
DPM	0.0345548	0.553897

Example 4.5 (Continued)

Table 4.5 compares the Z indices for the medical device diameters based on short-term and long-term estimates of sigma. The short-term standard deviation was obtained using the average moving range. Notice that the estimated short-term standard deviation is about 10% less than the long-term estimate, resulting in a lower estimated value for DPM. This is not surprising, since consistent product is usually easier to produce over a short period of time than over a longer period.

4.5.2 C_p and P_p

Another index commonly used to measure process capability is the index C_p, defined by

$$C_P = \frac{USL - LSL}{6\sigma} \tag{4.29}$$

C_p divides the distance between the specification limits by 6 times the standard deviation. As shown in Figure 4.7, this index can be thought of as the "design tolerance" divided by the "natural tolerance", measuring how much wider the allowable design tolerance is relative to the natural process variation. C_p can only be calculated when there is both an upper specification limit *USL* and a lower specification limit *LSL*.

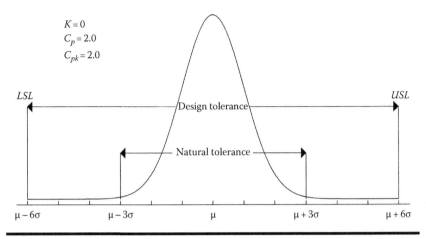

Figure 4.7   Capability indices for centered process.

A necessary but not sufficient condition for a process to be capable of satisfying the specification limits is that the design tolerance be greater than the natural tolerance of the process, that is, C_p be greater than 1. As will be demonstrated, such a requirement is not however a sufficient condition to guarantee conformance, since it assumes that the mean of the process is near the center of the design tolerance limits. If the process is not properly centered, the value of C_p may appear to be acceptable even though too much product is beyond the specification limits.

When estimating capability, it is common practice to label the index C_p when σ is replaced by its short-term estimate and P_p when σ is replaced by its long-term estimate. Such notation distinguishes well between *capability* and *performance*, but it is not universal. One must be careful when examining capability indices to be certain exactly what estimate of sigma they are based on.

Approximate confidence intervals and bounds may be constructed for C_p or P_p when the data are assumed to come from a normal distribution. The $100(1 - \alpha)\%$ confidence interval is

$$\hat{C}_P \sqrt{\frac{\chi^2_{1-\alpha/2,v}}{v}} \le C_P \le \hat{C}_P \sqrt{\frac{\chi^2_{\alpha/2,v}}{v}} \qquad (4.30)$$

where $\chi^2_{p,\upsilon}$ is the value of the chi-square distribution with υ degrees of freedom that is exceeded with probability p. υ is the degrees of freedom associated with the estimate of sigma used to compute the capability index. A one-sided lower bound for C_p is given by

$$\hat{C}_p \sqrt{\frac{\chi^2_{1-\alpha,\upsilon}}{\upsilon}} \leq C_p \tag{4.31}$$

Table 4.6 shows the degrees of freedom associated with various estimates of sigma.

Table 4.6   Degrees of Freedom Associated with Various Estimates of Sigma

Estimate	Equations	Degrees of Freedom υ
Long-term	(4.8) and (4.9)	$n - 1$
Short-term using pooled standard deviation	(4.11) and (4.12)	$\sum\limits_{j=1}^{m} (n_j - 1)$
Short-term using average subgroup range	(4.13)	$0.9 \sum\limits_{j=1}^{m} (n_j - 1)$
Short-term using average subgroup standard deviation	(4.15) and (4.16)	$0.88 \sum\limits_{j=1}^{m} (n_j - 1)$ to $\sum\limits_{j=1}^{m} (n_j - 1)$ depending on n
Short-term using average moving range	(4.18)	$n - 1$
Short-term using median moving range	(4.19)	$n - 1$
Short-term using squared successive differences	(4.20) and (4.21)	$n - 1$

Different rules of thumb exist for how large C_p and P_p need to be in order for process quality to be acceptable; in other words, how much wider than the natural tolerance should the design tolerance be? Commonly quoted minimum values are 1.33, 1.5, and 2.0, depending on whether you are qualifying a new process or monitoring an existing process. Clearly, the acceptable level depends on the corresponding proportion of nonconforming items. That proportion in turn depends on how far the process mean is from the specification limits, which is not accounted for by C_p.

The developers of Six Sigma observed many real-life processes and noted that, in most of them, the short-term mean was not constant. In fact, it was common for the short-term mean to fluctuate around the long-term mean by as much as 1.5σ. To examine the effect of such fluctuations on the proportion of nonconforming items, let K equal the offset between the process mean and the target value T as a multiple of the standard deviation:

$$K = \frac{\mu - T}{\sigma} \tag{4.32}$$

Assuming that the specification limits are symmetrically spaced about the target value, Table 4.7 shows the number of nonconforming items per million for selected values of C_p and K.

Six Sigma practitioners suggest that C_p should be ≥ 2.0, that is, the process sigma be small enough that the specification limits are at least 12 standard deviations apart. In other words, the design tolerance should be at least twice as wide as the natural tolerance. If the process is properly centered and the process mean does not vary around its long-term value by more than 1.5σ, the *DPM* associated with such a "6-sigma" process will equal no more than 3.4 defects per

Table 4.7   DPM for Different Combinations of C_p and K

C_p	$K = 0$	$K = 0.5$	$K = 1.0$	$K = 1.5$	$K = 2.0$
0.333	317,309	375,343	522,753	697,674	842,695
0.667	45,500	73,017	160,005	308,769	500,035
1.000	2,700	6,442	22,782	66,810	158,655
1.333	63.37	236.1	1,350	6,210	22,750
1.667	0.57	3.42	31.69	232.7	1,350
2.000	0.00	0.02	0.29	3.40	31.69

million, corresponding to the tabled value for C_p = 2.0 and K = 1.5. If the mean remains closer to the target than 1.5 times the standard deviation or if a *DPM* of greater than 3.4 is acceptable, then C_p could be less than 2.0 and still yield acceptable product.

Example 4.5 (Continued)

Table 4.8 displays the calculated values of C_p and P_p for the medical device diameters using the estimated sigma values in Table 4.5, together with lower 95% confidence bounds. It may be stated with 95% confidence that the capability indices of the medical device production process satisfy $C_p \geq 1.811$ and $P_p \geq 1.636$, assuming that the data come from a normal distribution.

4.5.3 C_r and P_r

An alternative to C_p, which is also used when both upper and lower specification limits are present, is the capability ratio defined by

$$C_r = \frac{6\sigma}{USL - LSL} \tag{4.33}$$

Table 4.8   Estimated Capability and Performance Indices for Medical Device Diameters

Capability Indices for Diameter			
Specifications: $LSL = 1.9$, $Nom = 2.0$, $USL = 2.1$			
	Short-Term Capability		Long-Term Performance
Sigma	0.016235		0.0179749
C_p and P_p	2.05317		1.85444
C_{pk} and P_{pk}	1.79796		1.62393
Based on 6.0 sigma limits. Short-term sigma estimated from average moving range.			
95.0% Confidence Bounds			
Index	Lower Limit	Index	Lower Limit
C_p	1.81127	P_p	1.63595
C_{pk}	1.58076	P_{pk}	1.42634

C_r represents the proportion of the design tolerance that is covered by the natural tolerance of the process. Since C_r is just the reciprocal of C_p, the equivalent requirement to $C_p \geq 2.0$ is $C_r \leq 0.5$. Likewise, the confidence limits for C_p may be inverted to give confidence limits for C_r.

4.5.4 C_{pk} and P_{pk}

C_p and C_r have two major drawbacks:

1. They require both upper and lower specification limits and so cannot be calculated when dealing with one-sided specifications.

2. They do not include the process mean. If the mean is far from the target value, it is possible to get acceptable values for the two-sided C_p and C_r indices even though much of the product is out of spec.

For these reasons, an increasingly popular capability index is C_{pk}. To calculate C_{pk}, two one-sided indices are first calculated to measure the distance from the process mean to whichever specification limits are present:

$$C_{pk(lower)} = \frac{\mu - LSL}{3\sigma} \qquad (4.34)$$

$$C_{pk(upper)} = \frac{USL - \mu}{3\sigma} \qquad (4.35)$$

The combined index C_{pk} is then the smaller of the two indices (assuming both are calculated):

$$C_{pk} = \min\left[C_{pk(lower)}, C_{pk(upper)} \right] \qquad (4.36)$$

C_{pk} will always be less than or equal to C_p. The two indices are equal only if the sample mean is exactly halfway between the upper and lower specification limits. As with C_p, the index will be labeled either C_{pk} or P_{pk} depending on whether a short-term or long-term estimate of sigma is used to calculate it.

Figure 4.8 shows a normal distribution that is not centered between the specification limits. Instead, the mean has been shifted a distance of 1.5σ in the positive direction. While the value of C_p is the same as in Figure 4.7, C_{pk} has been reduced from 2.0 to 1.5.

Because it incorporates estimates of both the process mean and the standard deviation, C_{pk} is closely related to the

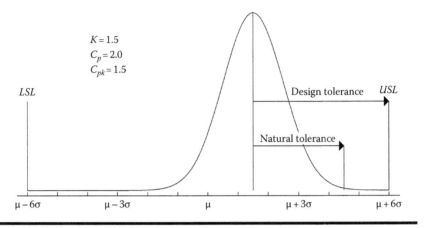

Figure 4.8 Capability indices for process with 1.5σ shift in mean.

percentage of the product that is beyond the specification limits. In fact, it is related to the Z index according to

$$C_{pk(lower)} = \frac{Z_{lower}}{3} \quad \text{and} \quad C_{pk(upper)} = \frac{Z_{upper}}{3} \qquad (4.37)$$

The proportion of items beyond the specification limits can thus be calculated from the C_{pk} values using

$$\theta = \Phi\left[-3C_{pk(lower)}\right] + 1 - \Phi\left[3C_{pk(upper)}\right] \qquad (4.38)$$

given both upper and lower specification limits. If either limit does not exist, the corresponding term is omitted. In many cases, either $C_{pk(upper)}$ or $C_{pk(lower)}$ will be very large because the process mean is far from the corresponding specification limit. In such cases, a good approximation for the estimated proportion beyond the specification limits is

$$\theta = 1 - \Phi\left[3C_{pk}\right] \qquad (4.39)$$

If the data come from a normal distribution, confidence limits can be calculated for C_{pk}. An approximate two-sided $100(1 - \alpha)\%$ confidence interval is given by

$$\hat{C}_{PK}\left[1 - Z_{\alpha/2}\sqrt{\frac{1}{9n\hat{C}_{PK}^2} + \frac{1}{2v}}\right] \leq C_{PK} \leq \hat{C}_{PK}\left[1 + Z_{\alpha/2}\sqrt{\frac{1}{9n\hat{C}_{PK}^2} + \frac{1}{2v}}\right]$$

$$(4.40)$$

where

  n is the total number of observations

  v is the degrees of freedom corresponding to the estimator used for sigma

A $100(1 - \alpha)\%$ lower confidence bound is given by

$$\hat{C}_{PK}\left[1 - Z_{\alpha}\sqrt{\frac{1}{9n\hat{C}_{PK}^2} + \frac{1}{2v}}\right] \leq C_{PK} \qquad (4.41)$$

Example 4.5 (Continued)

Table 4.8 includes estimates for both C_{pk} and P_{pk} using the medical device diameters. It may be stated with 95% confidence that $C_{pk} \geq 1.581$ and $P_{pk} \geq 1.426$.

4.5.5 C_m and P_m

C_m, the machine capability index, is very similar to C_p, except that the denominator is 8 times sigma rather than 6 times sigma:

$$C_m = \frac{USL - LSL}{8\sigma} \qquad (4.42)$$

For this index, the "natural tolerance" is defined as $\mu \pm 4\sigma$ rather than $\mu \pm 3\sigma$. All of the results given for C_p apply to C_m with the appropriate correction for the different denominator.

4.5.6 C_{pm}

Recall that one of the criticisms of C_p is that it does not account for the location of the process mean. Consequently, it can dramatically overestimate the quality of a process if the process mean is not close to the nominal or target value T. C_{pm} is a modified version of C_p that reduces its value using the difference between the process mean and T. It is defined by

$$C_{pm} = \frac{C_p}{\sqrt{1 + \frac{(\mu - T)^2}{\sigma^2}}} \qquad (4.43)$$

As the process mean diverges from the target value, C_{pm} becomes smaller.

An approximate $100(1 - \alpha)\%$ confidence interval for C_{pm} is given by

$$\hat{C}_{pm}\sqrt{\frac{\chi^2_{1-\alpha/2,v}}{v}} \le C_{pm} \le \hat{C}_{pm}\sqrt{\frac{\chi^2_{\alpha/2,v}}{v}} \qquad (4.44)$$

where

$$v = n\frac{(1 + \lambda)^2}{1 + 2\lambda} \qquad (4.45)$$

and

$$\lambda = \left(\frac{\hat{\mu} - T}{\hat{\sigma}}\right)^2 \qquad (4.46)$$

Although labeled C_{pm}, this index is a performance measure and is always calculated using a long-term estimate of σ.

A $100(1 - \alpha)\%$ lower confidence bound for C_{pm} is given by

$$\hat{C}_{pm} \sqrt{\frac{\chi^2_{1-\alpha,v}}{v}} \leq C_{pm} \qquad (4.47)$$

4.5.7 CC_{pk}

CC_{pk} is a modified version of C_{pk}, which replaces the population mean μ with the target value T:

$$CC_{pk} = \min\left(\frac{T - LSL}{3\sigma}, \frac{USL - T}{3\sigma}\right) \qquad (4.48)$$

It measures how capable the process is of meeting the specification if the sample mean was exactly at the target value. It is a capability measure only, calculated using a short-term estimate of σ.

4.5.8 *K*

Another popular capability index is K, which measures how much the population mean differs from the target value T. If the population mean is greater than or equal to the target, then

$$K = \frac{\mu - T}{USL - T} \qquad (4.49)$$

If the population mean is less than the target, then

$$K = \frac{\mu - T}{T - LSL} \qquad (4.50)$$

A value of K equal to 0.1 indicates that the population mean is located 10% of the way from the target to the upper specification limit. K is negative when the mean is less than the target value.

4.5.9 SQL: The Sigma Quality Level

Six Sigma practitioners quantify the quality level associated with a process by calculating

$$SQL = Z + 1.5 \qquad (4.51)$$

They refer to processes with an *SQL* of 6 or greater as having achieved "world class quality". Recall that Z refers to the number of standard deviations between the process mean and the nearer specification limit. Processes for which Z is never less than 4.5 have no more than 3.4 defects per million opportunities, since the tail area of the normal curve beyond 4.5 standard deviations equals 0.0000034. The additional 1.5 is added to Z based on the assertion that the short-term mean of most processes varies around its long-term value by no more than ±1.5 standard deviations. If this is true and the short-term Z is never less than 4.5, then the long-term mean must be at least 6 sigma removed from all specification limits. Hence the name "Six Sigma". Table 2.1 in Chapter 2 shows how *SQL* relates to other measurements of quality, assuming a 1.5 sigma shift.

Example 4.5 (Continued)

Table 4.9 shows a complete list of capability indices calculated from the medical device diameter data. The medical device production process is performing at a long-term sigma level equal to 6.37, which Six Sigma practitioners refer to as "world class quality". The rate of nonconforming items is less than 1 out of every million devices produced.

Table 4.10 shows confidence bounds for the quality indices, assuming that the data come from a normal distribution. For indices such as C_{pk} where larger values are better, one can state with 95% confidence that the true index is greater than or equal to the value shown. For indices such as C_r and K where smaller values are better, one can state with 95% confidence that the true index is less than or equal to the value shown.

Table 4.9   Capability and Performance Indices for Medical Device Diameters

Capability Indices for Diameter		
Specifications: $LSL = 1.9$, $Nom = 2.0$, $USL = 2.1$		
	Short-Term Capability	*Long-Term Performance*
Sigma	0.016235	0.0179749
C_p/P_p	2.05317	1.85444
C_r/P_r	48.7051	53.9246
C_m/P_m	1.53988	1.39083
Z_{upper}	6.92514	6.25484
Z_{lower}	5.39389	4.8718
Z_{min}	5.39389	4.8718
C_{pk}/P_{pk}	1.79796	1.62393
$C_{pk}/P_{pk(upper)}$	2.30838	2.08495
$C_{pk}/P_{pk(lower)}$	1.79796	1.62393
CC_{pk}	2.05317	
C_{pm}		1.52278
K		−0.1243
% beyond spec	0.00000345548	0.0000553897
DPM	0.0345548	0.553897
Sigma Quality Level	6.89	6.37

Note: Based on 6.0 sigma limits. Short-term sigma estimated from average moving range. The Sigma Quality Level includes a 1.5 sigma shift in the mean.

Table 4.10   Confidence Bounds for the Medical Device Quality Indices

95.0% Confidence Bounds			
Index	*Lower Quality Bound*	*Index*	*Lower Quality Bound*
C_p	1.81127	P_p	1.63595
C_r[a]	55.2098	P_r[a]	61.1264
C_m	1.35845	P_m	1.22697
Z_{upper}	6.09909	Z_{upper}	5.50541
Z_{lower}	4.74227	Z_{lower}	4.27903
Z_{min}	4.74227	Z_{min}	4.27903
C_{pk}	1.58076	P_{pk}	1.42634
$C_{pk(upper)}$	2.03303	$P_{pk(upper)}$	1.83514
$C_{pk(lower)}$	1.58076	$P_{pk(lower)}$	1.42634
CC_{pk}	1.81127		
		C_{pm}	1.35393
		K[a]	−0.0944546
% beyond spec[a]	0.000105851	% beyond spec[a]	0.000941031
DPM[a]	1.05851	*DPM*[a]	9.41031
Sigma Quality Level	6.24227	Sigma Quality Level	5.77903

[a] Lower quality bound corresponds to upper confidence bound for this index.

4.6 Confidence Bounds for Proportion of Nonconforming Items

The primary question posed in Chapter 1 was how to use a sample of data to estimate the proportion of items θ that do not conform to a set of specification limits. Having fit a normal distribution to a set of data and calculated $\hat{\theta}$, it is important to consider methods for obtaining confidence limits for θ.

4.6.1 Confidence Limits for One-Sided Specifications

It has been shown that θ is related to the capability index C_{pk} according to Equations 4.38 and 4.39. To obtain an upper confidence bound for θ when there is only one specification limit, the lower confidence bound for C_{pk} may be substituted directly into Equation 4.39.

> **Example 4.6 Confidence Limits for One-Sided Specifications**
>
> For the medical device diameters, the upper specification limit has little effect since $P_{pk(lower)}$ is much smaller than $P_{pk(upper)}$ and dominates the calculation of the proportion beyond the specification limits. Using Equation 4.41, the 95% lower confidence bound for P_{pk} equals
>
> $$1.62393 \left[1 - 1.645 \sqrt{\frac{1}{900(1.62393)^2} + \frac{1}{198}} \right] = 1.42634 \qquad (4.52)$$
>
> Equation 4.39 may then be used to create a 95% upper confidence bound for θ as
>
> $$\hat{\theta}_U = 1 - \Phi\left(3 * 1.42634\right) = 0.00000939 \qquad (4.53)$$
>
> The 95% upper bound on the proportion of nonconforming items is thus approximately 9.4 items per million.

4.6.2 Confidence Limits for Two-Sided Specifications

The problem is more difficult when there are two specification limits. One approach is to obtain a 95% upper bound for the proportion of items that exceed the upper spec and add that to the 95% upper bound for the proportion of items that are below the lower spec. Adding the two upper bounds gives a conservative estimate of the 95% upper bound for the overall proportion of defective items.

The approach outlined above tends to overestimate the upper bound for θ, since the estimated upper bounds of the two proportions are not independent. In particular, errors in the estimate of the process mean will cause a simultaneous underestimation of one proportion and overestimation of the other. Monte Carlo studies have shown that this overestimation is particularly significant when the mean is close to the target value (K is small) and sigma is large compared to the distance between the specification limits (C_p is small). In cases where the predominance of defects occur on one side of the mean rather than both, the overestimation effect is small.

Example 4.7 Confidence Limits for Two-Sided Specifications

For the medical device data, $P_{pk(lower)}$ = 1.62393 and $P_{pk(upper)}$ = 2.08495. The 95% lower confidence bounds for each of the one-sided P_{pk} values, calculated using Equation 4.41, are shown as follows:

Specification	Lower 95% Bound for P_{pk}	Upper 95% Bound for θ
Lower	1.42634	0.00000939
Upper	1.83514	0.00000002
Total		0.00000941

Also shown here are the corresponding upper 95% bounds for the proportion of items beyond each specification limit, calculated by substituting the lower bounds into Equation 4.39. The final 95% upper bound for the total θ equals the sum of the two bounds: $\hat{\theta}_{UCL}$ = 0.00000941. It may thus be claimed with 95% confidence that no more than 9.41 devices per million will be out of spec. Provided, of course, that the data are a random sample from a normal distribution.

An alternative approach for estimating an upper bound on θ when there are both upper and lower specification limits is to use bootstrapping. Bootstrap confidence limits are created by generating many subsamples from the available data and examining the distribution of the calculated indices among those subsamples.

4.6.2.1 Bootstrap Confidence Limits for Individuals Data

To generate bootstrap confidence limits when the data have been obtained as n individual observations, proceed as follows:

Step 1: Select a random sample of n observations from the data, sampling WITH REPLACEMENT. This implies that each time an observation is selected, all n observations have an equal probability of being selected, even those that have been selected previously.

Step 2: Calculate the sample mean $\bar{x}$ and estimate the long-term sigma.

Step 3: Select a random sample of $n - 1$ successive differences between consecutive observations in the original data, again sampling with replacement.

Step 4: Estimate the short-term sigma from the successive differences using the average moving range or another suitable method.

Step 5: Use the results from Steps 2 and 4 to estimate the quality indices and save them.

Step 6: Repeat Steps 2 through 5 a large number of times (50,000 times is common).

Step 7: Create a 95% upper or lower bound for each quality index by finding the 95th or 5th percentile of the values saved in Step 5. To create two-sided 95% confidence limits, find the 2.5th and 97.5th percentiles.

Example 4.7 (Continued)

Table 4.11 shows bootstrap estimates of the 95% confidence bounds for the proportion of nonconforming medical devices and for other quality indices. Note that the estimated upper bound for the long-term DPM is 9.00, which is slightly smaller than that obtained previously by summing the two upper bounds. A larger difference between the estimation methods would be expected if nonconforming items were not predominantly located beyond only one of the specification limits.

Table 4.11   Confidence Bounds for Capability Indices of Medical Device Diameters Obtained Using Bootstrapping

95.0% Confidence Bounds: Bootstrap Method (50,000 Subsamples)			
Index	Lower Quality Limit	Index	Lower Quality Limit
C_p	1.79826	P_p	1.60527
C_r[a]	55.6093	P_r[a]	62.2946
C_m	1.3487	P_m	1.20396
Z_{usl}	6.05081	Z_{usl}	5.32353
Z_{lsl}	4.70271	Z_{lsl}	4.28984
Z_{min}	4.70271	Z_{min}	4.28984
C_{pk}	1.56757	P_{pk}	1.42995
$C_{pk(upper)}$	2.01694	$P_{pk(upper)}$	1.77451
$C_{pk(lower)}$	1.56757	$P_{pk(lower)}$	1.42995
		C_{pm}	1.39957
		K[a]	−0.0945
% beyond spec[a]	0.000128413	% beyond spec[a]	0.000900381
DPM[a]	1.28413	DPM[a]	9.00381
Sigma Quality Level	6.20271	Sigma Quality Level	5.78984

[a] Lower quality bound corresponds to upper confidence bound for this index.

4.6.2.2 Bootstrap Confidence Limits for Subgroup Data

If the n observations are divided into m subgroups, each having size n_j, the approach described may still be used to estimate the mean and the long-term sigma. To estimate the short-term or within-group sigma, subsamples must be created that have the same structure as the original data. The jth subgroup of each bootstrap subsample should thus contain n_j deviations randomly sampled from the n deviations of the observations from their respective subgroup means. The within-group sigma may then be estimated using the desired method (such as the average subgroup range or standard deviation).

4.7 Summary

Given the assumption that data come from a normal distribution, estimates can be obtained of the proportion of nonconforming items using the methods described in this chapter. Bounds on that proportion may also be calculated. Related indices such as C_{pk} and P_{pk} may also be calculated to summarize the short-term capability and long-term performance of the process.

When the data do not come from a normal distribution, the calculated values can be misleading. As discussed in the next chapter, tests for normality should be performed before the methods described in this chapter are used. If the assumption of normality is not tenable, one of the methods described in Chapter 5 should be used instead.

Reference

Montgomery, D.C. (2013), *Introduction to Statistical Quality Control*, 7th edn., Hoboken, NJ: John Wiley & Sons.

Bibliography

Bissell, A.F. (1990), How reliable is your capability index? *Journal of the Royal Statistical Society, Series C*, **39**, 331–340.

Chan, L.K., Chen, S.W., and Spring, F. (1988), A new measure of process capability: C_{pm}, *Journal of Quality Technology*, **20**, 162–175.

Chernick, M.R. (1999), *Bootstrap Methods: A Practitioner's Guide*, New York: John Wiley & Sons.

Chou, Y.M., Owen, D.B., and Borrego, A.S.A. (1990), Lower confidence limits on process capability indices, *Journal of Quality Technology*, **22**, 223–229.

Johnson, N.L. and Kotz, S. (1993), *Process Capability Indices*, London, U.K.: Chapman & Hall.

Kotz, S. and Johnson, N.L. (2002), Process capability indices—A review, 1992–2000, *Journal of Quality Technology*, **34**, 2–53.

Kotz, S. and Lovelace, C.R. (1998), *Process Capability Indices in Theory and Practice*, London, U.K.: Arnold.

Kushler, R.H. and Hurley, P. (1992), Confidence bounds for capability indices, *Journal of Quality Technology*, **24**, 188–195.

Yum, B.J. and Kim, K.W. (2011), A Bibliography of the literature on process capability indices: 2000–2009, *Quality and Reliability Engineering International*, **27**, 251–268.

Chapter 5

Capability Analysis
of Nonnormal Data

The development of capability indices described in Chapter 4 assumes that the measured data values are a random sample from a normal distribution. Although the indices can still be calculated as defined there for data from other distributions, their meaning is not the same. In particular, the correspondence between specific values of indices such as C_{pk} and the proportion of nonconforming items does not hold if the distribution is not normal. Likewise, the confidence intervals do not provide the stated level of confidence.

To perform a capability analysis when the assumption of normality does not hold, the analyst has four choices:

1. Use the methods based on counting the number of non-conformities described in Chapters 2 and 3. However, since this approach tabulates only whether items are in spec or out of spec, it throws away a lot of information by not accounting for how close or far from the specification limits the measurements are. Consequently, large

sample sizes are needed to obtain a reliable estimate of
the proportion of nonconforming items.

2. Seek a transformation of the data such that after the
   transformation, the normal distribution is appropriate. This approach allows all of the results presented in
   Chapter 4 to be applied to the transformed data.

3. Find an alternative distribution that fits the data well. This
   requires modifying the indices so that they maintain the
   same relationship to *DPM* as when the normal distribution is appropriate.

4. Select a Johnson curve that matches the first four
   moments of the observed data and calculate nonnormal
   capability indices based on that curve.

This chapter describes the latter three approaches, after first
describing tests designed to indicate whether or not the
assumption of normality is tenable.

5.1 Tests for Normality

In order to decide whether the results of Chapter 4 can reasonably be applied to a data sample, it is necessary to determine whether or not the data could reasonably have come
from a normal distribution. This requires conducting a statistical test of the following hypotheses:

> **Null hypothesis H₀:** data come from a normal distribution
> **Alternative hypothesis Hₐ:** data do not come from a nor-
> mal distribution

Since in statistical tests the benefit of the doubt is given to the
null hypothesis, such a test determines whether or not there
is sufficient evidence to reject the hypothesis of normality.
Passing the test is not proof of normality, but failing it indicates that another approach should be used.

The first step in testing normality is the creation of a *quantile-quantile* or *Q-Q plot*. A Q-Q plot compares the order statistics of the data sample to the equivalent quantiles of a normal distribution having the same mean and standard deviation as the data. A line is superimposed on the plot to help determine whether or not the quantiles of the data correspond closely enough to the expected values for observations from a normal distribution.

Example 5.1 Tests of Normality

Figure 5.1 shows a Q-Q plot for the medical device data. The sorted values of diameter $x_{(i)}$ are plotted on the vertical axis. The corresponding positions on the horizontal axis are the percentiles of a normal distribution with the same mean and standard deviation as the data, evaluated at

$$p_i = \frac{i - 0.3175}{n + 0.365}, \quad i = 1, 2, \ldots, n \quad (5.1)$$

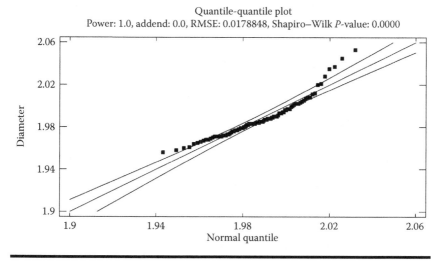

Quantile-quantile plot
Power: 1.0, addend: 0.0, RMSE: 0.0178848, Shapiro–Wilk *P*-value: 0.0000

Figure 5.1   Normal quantile-quantile (Q-Q) plot for untransformed medical device diameters.

95% probability limits for the percentiles of the fitted normal distribution are plotted on either side of the diagonal line to help judge whether or not the data are close enough to have come from a normal distribution. The medical device data show pronounced curvature, typical of data with a shorter lower tail and longer upper tail than data sampled from a normal distribution.

There are many tests that may be used to determine whether data come from a normal distribution. For sample sizes $n \leq 5000$, there is evidence that the most powerful test is one developed by Shapiro and Wilk (1965). Their test constructs a statistic W, which is roughly based on how close the points lie to a straight line on the quantile-quantile plot. Associated with W is a P-value that may be used to test the hypotheses stated earlier. A P-value less than α, the significance level of the test, leads to rejection of the null hypothesis.

Example 5.1 (Continued)

The quantile-quantile plot for the medical device data in Figure 5.1 displays the P-value of the Shapiro-Wilk test. The P-value is well below $\alpha = 0.01$, leading to rejection of the hypothesis of normality at the 1% significance level.

For data sets in which $n > 5000$, an alternative test such as the Anderson-Darling goodness-of-fit test may be used. The Anderson-Darling test measures the area between the empirical and fitted cumulative distribution functions. As originally derived, P-values used to test goodness-of-fit using the A-D test assume that the parameters of the distribution are known (not a likely assumption when fitting a distribution to data). Thankfully, special modifications of the test have been derived, which allow it to be used when the parameters of the normal distribution have been estimated from the data, as discussed in D'Agostino and Stephens (1986). The Anderson-Darling test is discussed further in Section 5.3.2.

5.2 Power Transformations

When the data are such that the hypothesis of normality is rejected, it is often possible to find a transformation of the data such that the normal distribution is appropriate in the transformed metric. If such a transformation can be found, then the calculations presented in Chapter 4 can be applied to the transformed data.

A widely used class of transformations transforms the random variable X according to

$$X' = (X + \Delta)^p \quad \text{for } p \neq 0 \tag{5.2}$$

In this transformation, the variable X is raised to the power p, after adding an addend equal to Δ. To make the transformation a continuous function of p, the natural logarithm is used when $p = 0$:

$$X' = \ln(X + \Delta) \quad \text{for } p = 0 \tag{5.3}$$

Since these power transformations are nonlinear, the distribution of the transformed data has a different shape than the distribution of the original values. If $p < 1$, the lower tail of the distribution is lengthened while the upper tail is shortened. If $p > 1$, the reverse is true. The addend Δ is often set equal to 0. However, it must be large enough that all values of $(X + \Delta)$ are positive.

The general class of power transformations defined by (5.2) and (5.3) includes several special cases, as listed in Table 5.1. The farther p is from 1 in either direction, the stronger the effect of the transformation on the shape of the distribution.

Table 5.1   Common Power Transformations

Power	Interpretation
$p = -2$	Reciprocal square
$p = -1$	Reciprocal
$p = -0.5$	Reciprocal square root
$p = -1/3$	Reciprocal cube root
$p = 0$	Logarithm
$p = 1/3$	Cube root
$p = 0.5$	Square root
$p = 1$	Original data
$p = 2$	Square
$p = 3$	Cube

5.2.1 Box-Cox Transformations

When a transformation such as that defined earlier is applied to X, it changes not only the shape of the distribution but also the magnitude of the data values. This makes it hard to compare directly the effect of applying different powers. To avoid that problem, Box and Cox (1964) suggested a modification of the transformation that preserves the magnitude of the data while changing only the shape. They proposed the following transformation:

$$X' = \frac{\left(X + \lambda_2\right)^{\lambda_1} - 1}{\lambda_1 g^{\lambda_1 - 1}} \quad \text{if } \lambda_1 \neq 0 \tag{5.4}$$

$$X' = g \ln\left(X + \lambda_2\right) \quad \text{if } \lambda_1 = 0 \tag{5.5}$$

In their transformation, λ_1 is the power, λ_2 is the addend, and g is the geometric mean of $(X + \lambda_2)$. The effect of the

Box-Cox transformation on the shape of the distribution is equivalent to that proposed in the previous section if $\lambda_1 = p$ and $\lambda_2 = \Delta$.

To find the optimal values of λ_1 and λ_2, Box and Cox proposed minimizing the mean squared error of the transformed data:

$$MSE = \frac{\sum_{i=1}^{n} \left(x_i' - \bar{x}' \right)^2}{n} \tag{5.6}$$

Numerical methods are commonly used to find the optimal values of λ_1 and λ_2.

Example 5.2 Power Transformations

When applying the Box-Cox procedure to data in which all of the values are positive, it is common practice to start by setting $\Delta = 0$. For the medical device diameters, the value of p that minimizes the MSE when the range of powers is restricted to $p = -5$ to $+5$ is $p = -5$. As may be seen from Figure 5.2, the transformation has little effect on the shape

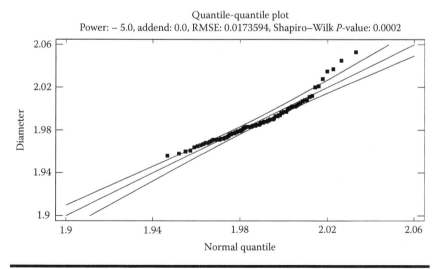

Figure 5.2   Q-Q plot after optimizing only the power.

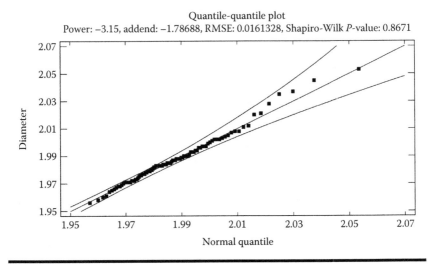

Figure 5.3   Q-Q plot after optimizing both the power and addend.

of the distribution. The fact that the best value of p is at one end of the allowable range is an indication that the power transformation does not work well with an addend equal to 0. This is often the case when the coefficient of variation of the data is small.

Figure 5.3 shows the results obtained when both the power and the addend are optimized. The resulting power transformation is

$$X' = \left(diameter - 1.78688\right)^{-3.15} \qquad (5.7)$$

The Shapiro-Wilk P-value is now well above 0.1, indicating that there is no reason to reject the hypothesis of normality for the transformed data. The points also lie much closer to the diagonal line on the Q-Q plot. The reason that the transformation works well once the addend is allowed to be nonzero is that it shifts the origin of the transformed data to a positive value 1.78688, thereby increasing the coefficient of variation and allowing the power to have a greater impact on the shape of the distribution.

Note: Many distributions that are useful for modeling continuous data such as the lognormal and gamma distributions are only defined for values of $X > \tau$, where τ is called the "lower threshold" of the distribution. If $\tau = 0$, as is often the case, then the distribution is defined for all positive values of X. When fitting a power transformation, the values of X are restricted such that $X + \Delta > 0$. Consequently, the implied lower threshold of the distribution when estimating a power transformation involving a nonzero addend is $\tau = -\Delta$.

5.2.2 Calculating Process Capability

After applying the power transformation, the transformed data usually have no direct interpretation. What the transformation does is provide a metric in which the variability among the data values is well approximately by a normal distribution and in which the results of procedures that assume normality can be applied. This includes all of the methods described in Chapter 4 for calculating capability indices and the proportion of nonconforming items.

Example 5.3 Calculating Process Capability for Transformed Data

To calculate process capability for the medical device data, the entire problem (both the data and the specification limits) must first be transformed using Equation 5.7. The transformed specification limits are given by

$$LSL' = \left(1.9 - 1.78688\right)^{-3.15} = 957.968 \qquad (5.8)$$

and

$$USL' = \left(2.1 - 1.78688\right)^{-3.15} = 38.7714 \qquad (5.9)$$

Because of the negative power, the limits have switched from lower to upper and vice versa. Performing a capability analysis on the transformed data with the transformed

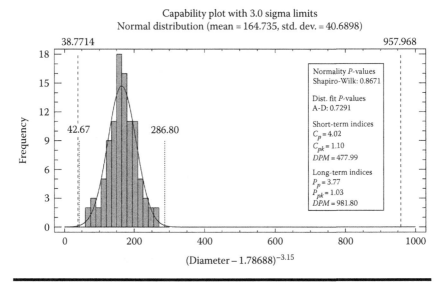

Figure 5.4 Process capability analysis for the transformed medical device data.

specification limits yields the results shown in Figure 5.4. Notice that the transformed data are much more symmetric than the original data and appear to be well-described by a normal distribution. The Shapiro-Wilk *P*-value is well above 0.1, indicating that the transformed data are adequately characterized by a normal distribution.

Since there is no intrinsic meaning to the data when expressed in the transformed metric, it is often more appealing to apply an inverse transformation to the fitted normal distribution and plot everything in the original units. Figure 5.5 shows the same analysis as Figure 5.4, except that the normal distribution has been expressed in the original units of diameter. Note that the implied distribution in the original metric is positively skewed, which matches the previously noted properties of the data.

To calculate capability indices, the equations presented in Chapter 4 may now be applied to the transformed values of diameter. For example, the value of P_p would be calculated by dividing the difference between the

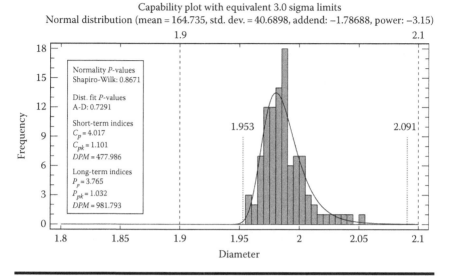

Figure 5.5 Inverse transformation after fitting normal distribution in transformed metric.

transformed specification limits by 6 times the standard deviation of the transformed data.

Example 5.3 (Continued)

The value of P_p based on the transformed data is

$$P_p = \frac{957.968 - 38.7714}{6 \cdot 40.6898} = 3.765 \qquad (5.10)$$

Compared to the results given in Chapter 4, which assumed that the original data came from a normal distribution and gave an estimate of $P_p = 1.85$, this is a fairly dramatic increase. On the other hand, P_{pk} has decreased from 1.62 before the transformation to 1.03 after the transformation. This results in a corresponding increase in the long-term *DPM* from less than 1 to almost 982. The long upper tail that was visible in the data when drawn in the untransformed metric is now a much shorter lower tail in the transformed metric. However, the transformed mean is much closer to the transformed specification limit, resulting in a larger value of *DPM*.

5.2.3 Confidence Limits for Capability Indices

To calculate confidence limits for the capability indices, the results of Chapter 4 are applied in the transformed metric. Any assumptions of normality apply in that metric, rather than in the original untransformed units.

> **Example 5.3 (Continued)**
>
> Table 5.2 shows the estimated values for various capability indices, all calculated using the transformed medical device diameters. Table 5.3 shows 95% lower confidence bounds for the indices. Based on the analysis, it may be stated with 95% confidence that $P_{pk} \geq 0.899$.

It should be noted that two methods have been suggested in the literature for calculating capability indices when transforming data by a power transformation:

1. Transforming both the data and the specification limits and calculating the indices in the transformed metric. This is the approach that has been used here.
2. Transforming the data, finding percentiles of the normal distribution in the transformed metric that correspond to $\mu \pm 3\sigma$, inversely transforming the percentiles, and then calculating the indices using the distance between the percentiles instead of 6σ.

The major advantage of the first method is that it preserves the relationship between the indices and the proportion of nonconforming items. Using the first approach, $C_{pk} = 1.5$ continues to imply that $DPM = 3.4$, which would not be true using the second approach.

Table 5.2   Capability Indices Calculated in the Transformed Metric

	Short-Term Capability	*Long-Term Performance*
Sigma (after transformation)	38.1337	40.6898
C_p/P_p	4.01743	3.76506
C_r/P_r	24.8915	26.56
C_m/P_m	3.01307	2.8238
Z_{upper}	3.30321	3.09571
Z_{lower}	20.8014	19.4947
Z_{min}	3.30321	3.09571
C_{pk}/P_{pk}	1.10107	1.0319
$C_{pk}/P_{pk(upper)}$	1.10107	1.0319
$C_{pk}/P_{pk(lower)}$	6.93379	6.49822
CC_{pk}	−7.23508	
C_{pm}		2.86682
K		−0.0416431
% beyond spec	0.0477986	0.0981793
DPM	477.986	981.793
Sigma Quality Level	4.8	4.6

Note: Based on 6.0 sigma limits in the transformed metric. Short-term sigma estimated from average moving range. The Sigma Quality Level includes a 1.5 sigma shift in the mean.

5.3 Fitting Alternative Distributions

Many probability distributions have been developed over the years for modeling data. When the normal distribution does not fit well, an alternative to transforming the data is to search for a different distribution that does a better job of describing

Table 5.3 95% Confidence Bounds for Capability Indices Calculated in the Transformed Metric

Index	Lower Quality Limit	Index	Lower Quality Limit
C_p	3.54411	P_p	3.32147
C_r[a]	28.2159	P_r[a]	30.1071
C_m	2.65808	P_m	2.4911
Z_{upper}	2.8835	Z_{upper}	2.6982
Z_{lower}	18.3642	Z_{lower}	17.2099
Z_{min}	2.8835	Z_{min}	2.6982
C_{pk}	0.961168	P_{pk}	0.899401
$C_{pk(upper)}$	0.961168	$P_{pk(upper)}$	0.899401
$C_{pk(lower)}$	6.12141	$P_{pk(lower)}$	5.73664
CC_{pk}	0.0		
		C_{pm}	2.56168
		K[a]	−0.302878
% beyond spec[a]	0.253075	% beyond spec[a]	0.437011
DPM[a]	1966.45	DPM[a]	3485.77
Sigma Quality Level	4.3835	Sigma Quality Level	4.1982

[a] Lower quality bound corresponds to upper confidence bound for this index.

the observed data. Often, a distribution can be found that has the same basic shape as the population from which the sample was taken.

Table 5.4 lists 27 distributions that may be used to model continuous data. Each distribution has between 1 and 3 parameters. Note that some distributions have restrictions on the range of values for X. Further information about each distribution may be found in Appendix A.

Table 5.4   Distributions for Continuous Data

Distribution	Parameters	Range of X
Birnbaum-Saunders	Shape β, scale θ > 0	X > 0
Cauchy	Mode θ, scale β > 0	All real X
Exponential	Rate λ > 0	X ≥ 0
Exponential (2-parameter)	Threshold θ, scale λ > 0	X ≥ θ
Exponential power	Mean μ, scale β ≥ –1, scale φ > 0	All real X
Folded normal	Location μ > 0, scale σ ≥ 0	X ≥ 0
Gamma	Shape α > 0, scale λ > 0	X ≥ 0
Gamma (3-parameter)	Threshold θ, shape α > 0, scale λ > 0	X ≥ θ
Generalized gamma	Location μ, scale σ > 0, shape λ > 0	X > 0
Generalized logistic	Location μ, scale κ > 0, shape γ > 0	All real X
Half normal	Threshold μ, scale σ > 0	X ≥ μ
Inverse Gaussian	Mean θ > 0, scale β > 0	X > 0
Laplace	Mean μ, scale λ > 0	All real X
Largest extreme value	Mode γ > 0, scale β > 0	All real X
Logistic	Mean μ, std. dev. σ > 0	All real X
Loglogistic	Median exp(μ), shape σ > 0	X > 0
Loglogistic (3-parameter)	Threshold θ, median exp(μ), shape σ > 0	X > θ
Lognormal	Location μ, scale σ > 0	X > 0
Lognormal (3-parameter)	Threshold θ, location μ, scale σ > 0	X > θ

(*Continued*)

Table 5.4 (*Continued*) Distributions for Continuous Data

Distribution	Parameters	Range of X
Maxwell	Threshold θ, scale β > 0	$X > \theta$
Normal	Mean μ, std. dev. σ > 0	All real X
Pareto	Shape $c > 0$	$X \geq 1$
Pareto (2-parameter)	Threshold θ > 0, shape $c > 0$	$X \geq \theta$
Rayleigh	Threshold θ, scale β > 0	$X > \theta$
Smallest extreme value	Mode γ > 0, scale β > 0	All real X
Weibull	Shape α > 0, scale β > 0	$X \geq 0$
Weibull (3-parameter)	Threshold θ, shape α > 0, scale β > 0	$X \geq \theta$

5.3.1 Selecting a Distribution

If no theory exists to explain why a random variable should follow a particular distribution, the data themselves must be used to suggest a reasonable model. A useful way to proceed is to fit all of the distributions and rank their goodness-of-fit from best to worst according to one of various measures:

1. *Log likelihood*: calculates the natural logarithm of the likelihood function for each distribution

$$\ln\left[L\left(\hat{\theta}\right)\right] = \sum_{i=1}^{n} \ln[\hat{f}\left(x_i | \hat{\theta}\right)] \qquad (5.11)$$

where $\hat{f}\left(x_i | \hat{\theta}\right)$ is the probability density function fitted to the data for that distribution. Larger values of the log likelihood function usually indicate better fitting distributions.

2. *Kolmogorov-Smirnov D*: calculates the maximum distance between the cumulative distribution function (CDF) of the data and the CDF of the fitted distribution. Smaller values of D correspond to better fits.
3. *Anderson-Darling A^2*: calculates a weighted measure of the area between the empirical and fitted CDFs. Smaller values of A^2 correspond to better fits.

Example 5.4 Fitting an Alternative Distribution

Table 5.5 shows statistics derived when each of the 27 distributions is fit to the medical device diameters. The distributions have been sorted according to the value of the Anderson-Darling statistic. There is some disagreement among the goodness-of-fit statistics as to which distribution provides the best fit. The log likelihood statistic and the Anderson-Darling statistic give the best results when using the 3-parameter loglogistic distribution. The Kolmogorov-Smirnov D statistic is smallest for the generalized logistic distribution. Among the distributions with no more than 2 parameters, the largest extreme value distribution is the best according to all of the criteria.

In choosing between alternative distributions, it is important to consider several factors:

1. In general, simpler models (those with fewer parameters) are preferable to more complicated models. Adding a third parameter to a distribution will always allow it to conform more closely to an observed set of data. But a third parameter increases the risk of overfitting the sample data, which may in the long run give a poorer description of the population from which the data came. If two distributions perform similarly and one has less parameters than the other, the distribution with less parameters will usually be preferable.

Table 5.5 Goodness-of-Fit Statistics for Distributions Fit to the Medical Device Diameters

Comparison of Alternative Distributions

Distribution	Est. Parameters	Log Likelihood	KS D	A^2
Loglogistic (3-parameter)	3	271.122	0.0633437	0.178586
Generalized logistic	3	270.752	0.0627998	0.280921
Largest extreme value	2	270.696	0.0670121	0.309344
Lognormal (3-parameter)	3	270.492	0.0652655	0.347718
Gamma (3-parameter)	3	270.023	0.0677877	0.452564
Exponential power	3	266.151	0.0579234	0.777642
Loglogistic	2	265.245	0.0693762	0.844609
Rayleigh	2	268.228	0.0892246	0.859113
Logistic	2	264.939	0.0703107	0.881328
Weibull (3-parameter)	3	268.338	0.081651	0.882886
Laplace	2	266.126	0.0614207	0.925591
Maxwell	2	267.356	0.103046	0.975576
Cauchy	2	257.124	0.0942539	1.37669
Generalized gamma	3	261.023	0.123932	1.83336

(Continued)

Table 5.5 (*Continued*) Goodness-of-Fit Statistics for Distributions Fit to the Medical Device Diameters

Comparison of Alternative Distributions

Distribution	Est. Parameters	Log Likelihood	KS D	A^2
Inverse Gaussian	2	261.022	0.123939	1.83356
Birnbaum-Saunders	2	261.022	0.123939	1.83356
Lognormal	2	261.02	0.124211	1.85076
Gamma	2	260.846	0.124611	1.86193
Folded normal	2	260.487	0.12596	1.91975
Normal	2	260.484	0.126229	1.93765
Half normal	2	259.025	0.206249	5.99829
Smallest extreme value	2	234.823	0.16621	6.71728
Pareto	1	−131.127	0.623466	44.7816
Exponential	1	−168.691	0.626231	45.1176
Exponential (2-parameter)	2	245.555	0.258199	—
Weibull	2	161.746	0.190599	—
Pareto (2-parameter)	2	245.011	0.260184	—

2. Certain 3-parameter distributions such as the 3-parameter loglogistic and the 3-parameter lognormal distribution add an additional threshold parameter to distributions that are normally defined for $X > 0$. This is similar to using a nonzero addend when transforming the data in the previous section. As will be seen in the following example, adding a lower threshold can cause problems when estimating the lower tail of the distribution.

Example 5.4 (Continued)

Figure 5.6 shows the 3 best-fitting distributions for the medical device data. The 2-parameter largest extreme value distribution and the 3-parameter generalized logistic distribution are almost identical. The estimated parameters are shown in Table 5.6.

Consider for a moment the 3-parameter loglogistic distribution. Allowing for a third parameter shifts the lower limit of the distribution from $X = 0$ to $X = 1.93928$. Below that threshold, the distribution is not defined. In this case, it means that the distribution is not defined at the lower specification limit of 1.9, arbitrarily forcing the proportion of nonconforming items below that lower limit to be 0.

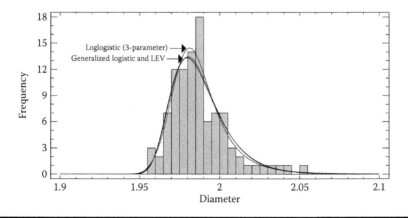

Figure 5.6   Best-fitting distributions for medical device diameters.

Table 5.6   Estimated Parameters for Distributions Fit to Medical Device Diameters

Fitted Distributions		
Generalized Logistic	*Largest Extreme Value*	*Loglogistic (3-Parameter)*
Location = 1.94242	Mode = 1.97962	Median = 0.0453747
Scale = 0.013316	Scale = 0.0138082	Shape = 0.198385
Shape = 17.0816		Lower threshold = 1.93928

Similar behavior occurs with the 3-parameter lognormal, gamma, and Weibull distributions. Selecting a distribution that is not defined at one of the specification limits is not good practice.

Given these considerations, the largest extreme value distribution appears to be the best alternative distribution for the medical device data. The fitted largest extreme value distribution is shown in Figure 5.7. Note that the distribution is characterized by its mode (the location at which the PDF

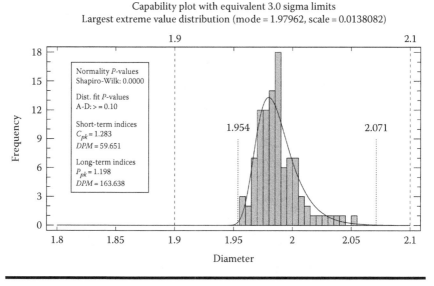

Figure 5.7   Largest extreme value distribution with equivalent "3-sigma" limits.

is highest) and a scale parameter that determines its spread. The short vertical lines on the plot are located at the 0.135% and 99.865% percentiles, which are equivalent to $\mu \pm 3\sigma$ when fitting a normal distribution.

5.3.2 Testing Goodness-of-Fit

The Kolmogorov-Smirnov (K-S) and Anderson-Darling (A-D) statistics may be used to test the hypothesis that a selected distribution provides a satisfactory model for data that were randomly sampled from a population. A typical set of hypotheses to be tested is

Null hypothesis H₀: data come from a largest extreme value distribution
Alternative hypothesis Hₐ: data do not come from a largest extreme value distribution

The K-S and A-D statistics compare the fitted cumulative distribution $\hat{F}(X)$ to the empirical distribution function (EDF) of the data. To perform the tests, the fitted distribution is first evaluated at each of the order statistics in the sample:

$$z_{(i)} = \hat{F}\left(x_{(i)}\right)$$

(5.12)

To calculate the K-S statistic, the maximum distances of the EDF above and below the fitted cumulative distribution are then calculated:

$$D^+ = \max_i \left\{ \frac{i}{n} - z_{(i)} \right\}$$

(5.13)

$$D^- = \max_i \left\{ z_{(i)} - \frac{i-1}{n} \right\}$$

(5.14)

The overall K-S statistic is the larger of the 2 distances:

$$D = \max\left(D^+, D^-\right) \tag{5.15}$$

The A-D statistic is based on the area between the fitted and empirical cumulative distribution functions. One formula for calculating it is

$$A^2 = -n - \frac{\sum_{i=1}^{n}\left((2i-1)\ln\left(z_{(i)}\right) + (2n+1-2i)\ln\left(1-z_{(i)}\right)\right)}{n}$$

$$\tag{5.16}$$

Unfortunately, it is not possible to calculate exact P-values for testing the stated hypotheses when the parameters of the distribution have been estimated from the data. However, modifications of the K-S and A-D statistics have been derived for many popular distributions and may be compared to tabulated critical values in order to decide whether or not to reject the null hypothesis. These modifications are discussed in detail in D'Agostino and Stephens (1986). In particular, to test the goodness-of-fit of the largest extreme value or Weibull distribution using the K-S test when the distribution parameters have been estimated from the data, the statistic $\sqrt{n}D$ is compared to a table of critical values derived from Monte Carlo studies. For a sample of $n = 100$ observations, the null hypothesis is rejected at the 5% significance level if $\sqrt{n}D$ is greater than 0.865 and at the 10% level if $\sqrt{n}D$ is greater than 0.795. Using the A-D statistic, A^2 is first converted to a modified value according to

$$A^2_{\mathrm{mod}} = A^2\left(1 + 0.2/\sqrt{n}\right) \tag{5.17}$$

If the modified value is greater than 0.757, the null hypothesis is rejected at the 5% significance level. If it is greater than 0.637, the null hypothesis is rejected at the 10% significance level.

Table 5.7 Goodness-of-Fit Tests for the Medical Device Diameters

Distribution: Largest Extreme Value		
	Kolmogorov-Smirnov D	*Anderson-Darling A²*
Statistic	0.0670121	0.309344
Modified form	0.670121	0.315531
P-value	≥0.10[a]	≥0.10[a]

[a] Indicates that the *P*-value has been compared to tables of critical values specially constructed for fitting the selected distribution.

Example 5.5 Testing Goodness-of-Fit of Nonnormal Distribution

Table 5.7 shows the results of performing the K-S and A-D tests on the medical device diameters after fitting a largest extreme value distribution. Both the original test statistics and the modified versions are shown. The *P*-values are both greater than or equal to 10%, implying that the largest extreme value distribution provides a reasonable model for the medical device data.

5.3.3 Calculating Capability Indices

When calculating capability indices for distributions other than the normal distribution, the formulas given in Chapter 4 cannot be used. Those formulas rely on modeling the data by a normal bell-shaped curve, which is completely specified by μ and σ. What must be done instead is as follows:

1. Use the fitted distribution $\hat{F}(x)$ to calculate $\hat{\theta}_{LSL}$ and $\hat{\theta}_{USL}$, the proportion of nonconforming items estimated to be beyond each specification limit.
2. Use the relationship between θ and Z given in Equation 4.28 to calculate *equivalent Z-indices*. An equivalent Z-index is the value of a standard normal random

variable that corresponds to the same estimated proportion of nonconforming items as the fitted distribution.
3. Use the relationship between Z and C_{pk}, SQL and other indices to generate equivalent values of those indices.

Example 5.6 Calculating Capability Indices for Nonnormal Distribution

Table 5.8 shows the results of fitting a largest extreme value distribution to the medical device diameters. It shows the following:

> *Equivalent 6.0 Sigma Limits*: Percentiles of the fitted largest extreme value distribution located at the same percentages as $\mu \pm 3\sigma$ when using a normal distribution. This equivalent "6-sigma" range, 1.95355 to 2.07085, contains 99.73% of the fitted distribution.
>
> *Observed Beyond Spec*: The percentages of the sample data below the lower specification limit and above the upper specification limit. Note that all of the data values are within the specification limits.
>
> *Estimated Beyond Spec*: The estimated probability of being below the lower specification limit and above the upper specification limit, based on the fitted largest extreme value distribution.
>
> *Z-Score*: Quantiles of the standard normal distribution corresponding to the estimated percentages beyond the specification limits.
>
> *Defects per Million*: The estimated percentages beyond the specification limits expressed as occurrences per million.

The fitted largest extreme value distribution has an estimated area of 0.016364% above the upper specification limit, corresponding to a $DPM = 163.64$. In the case of a standard normal distribution, that same area is found above $Z = 3.59$, implying that the upper specification limit is the equivalent of 3.59 standard deviations above the mean for a normal population. The Z-Score for the lower specification limit is too large to display, but corresponds to an estimated $DPM = 0$. The Z-Score for the target value

Table 5.8 Calculated *DPM* and Equivalent *Z*-Scores When Fitting a Largest Extreme Value Distribution to the Medical Device Diameters

Data variable: Diameter				
Transformation: None				
Distribution: Largest Extreme Value	Sample size = 100			
	Mean = 1.97962			
	Scale = 0.0138082			
	(Mean = 1.98759)			
	Std. dev. = 0.0177097			
Equivalent 6.0 Sigma Limits	99.865 percentile = 2.07085			
	Median = 1.98468			
	0.134996 percentile = 1.95355			
Specifications	*Observed Beyond Spec*	*Z-Score*	*Estimated Beyond Spec*	*Defects Per Million*
USL = 2.1	0.000000%	3.59	0.016364%	163.64
Nominal = 2.0		0.83		
LSL = 1.9	0.000000%		0.000000%	0.00
Total	0.000000%		0.016364%	163.64

is calculated by determining at what percentile of the fitted largest extreme value distribution that target value lies and converting it to a standard normal random variable. The conclusion derived from this analysis is that approximately 164 devices out of every million produced will have a diameter outside of the specification limits. This estimate is substantially above the estimate obtained when assuming normality in Chapter 4 but less than that obtained using the transformation approach described in Section 5.2.

Having obtained equivalent *Z*-indices using the fitted distribution, it is now possible to compute equivalent capability indices. Since the lower specification limit is so far down in the lower tail of the fitted distribution, the problem of estimating process capability for the medical device diameters is essentially one sided. To compute P_{pk}, insert the value of $Z_{min} = 3.59$ into Equation 4.37, which yields $P_{pk} = 3.59/3 = 1.2$. Similar calculations may be made for many of the other capability indices.

Table 5.9 shows a complete set of capability indices for the medical device diameters, based on the fitted largest extreme value distribution. Only those indices that are directly related to θ are displayed. All of the indices are derived from the equivalent Z-indices.

It will be noticed in Table 5.9 that estimates are given for both short-term capability and long-term performance.

Table 5.9 Calculated Capability Indices for the Medical Device Diameters Based on the Fitted Largest Extreme Value Distribution

Capability Indices for Diameter		
	Short-Term Capability	Long-Term Performance
Sigma (after normalization)	0.93374	1.0
Z_{upper}	3.84765	3.5927
Z_{min}	3.84765	3.5927
C_{pk}/P_{pk}	1.28255	1.19757
$C_{pk}/P_{pk(upper)}$	1.28255	1.19757
$C_{pk}/P_{pk(lower)}$		
CC_{pk}	0.98763	
K		−0.298612
% beyond spec	0.00596507	0.0163638
DPM	59.6507	163.638
Sigma Quality Level	5.35	5.09

Note: Based on 6.0 sigma limits in the normalized metric. Short-term sigma estimated from average moving range. The Sigma Quality Level includes a 1.5 sigma shift in the mean.

Fitting a distribution to the entire sample of n observations estimates the long-term performance of the process over the entire sampling period. When converted to equivalent Z-scores, the long-term sigma is standardized to a value of 1.

To calculate the equivalent short-term sigma, all observations are converted to Z-scores and the short-term and long-term sigmas are calculated from those scores in the usual manner. The equivalent short-term sigma is then set equal to the ratio of the calculated short-term and long-term sigmas and used to calculate the short-term indices.

5.3.4 Confidence Limits for Capability Indices

While the relationship between θ and Z can be used to calculate point estimates of equivalent capability indices, it cannot reliably be used to calculate confidence limits. This is because the properties of the estimated parameters of other distributions are not as well studied as for the normal distribution. One possible solution to this problem is to use bootstrapping. Bootstrapping is a method by which the sampling variability of a statistic can be determined by repeatedly sampling the available data.

In particular, suppose n observations have been sampled from an unknown population. To assess the sampling variability of a statistic such as P_{pk}, bootstrapping would proceed as follows:

1. Create another sample of size n by randomly selecting observations from the original sample *with replacement*. This means that after each observation is selected, it is returned to the original sample and may well be selected again.
2. Fit the selected distribution to the new sample.
3. Calculate an equivalent P_{pk} using the fitted distribution and store it somewhere.

4. Repeat Steps 1 through 3 a large number of times, perhaps 50,000 times. Using this technique accumulates 50,000 values for P_{pk}.
5. Calculate confidence limits and bounds on P_{pk} from the percentiles of the 50,000 calculated values. For example, the 5th percentile of the 50,000 values provides a one-sided lower 95% confidence bound for P_{pk}.

Example 5.6 (Continued)

Table 5.10 shows confidence bounds for several capability indices based on 50,000 subsets of the medical device data. With 95% confidence, it is estimated that the *DPM* is no greater than 455. Note: The "lower" quality bound displayed for *DPM* is actually its upper confidence bound, since high values of *DPM* are undesirable.

Table 5.10 Confidence Bounds for Quality Indices for the Largest Extreme Value Model Based on Bootstrapping

95.0% Confidence Bounds—Bootstrap Method (50,000 Subsamples)	
Index	Lower Quality Bound
Z_{upper}	3.317
Z_{min}	3.317
P_{pk}	1.10567
$P_{pk(upper)}$	1.10567
$P_{pk(lower)}$	6.14978
K[a]	−0.0373924
% beyond spec[a]	0.0454964
DPM[a]	454.964
Sigma Quality Level	4.817

[a] Lower quality bound corresponds to upper confidence bound for this index.

5.4 Nonnormal Capability Indices and Johnson Curves

Another method that has been suggested for estimating capability indices when the data do not conform to a normal distribution involves approximating the distribution of the data using Johnson curves. In 1949, N.L. Johnson developed a family of 4-parameter probability distributions that are capable of representing a population with any combination of skewness and kurtosis. The distributions are based on a set of normalizing transformations that convert the random variable X to a standard normal random variable Z according to

$$Z = \gamma + \delta f\left(\frac{X - \theta}{\lambda}\right) \qquad (5.18)$$

where
   θ is a location parameter
   λ is a scale parameter
   γ and δ are shape parameters
   $f(.)$ is a transformation function

   In order to match all combinations of possible values for the first four moments, three basic types of Johnson distributions with different transformation functions were defined:

1. *Unbounded distributions S_U*: These distributions are used for situations in which the standardized third moment $\sqrt{\beta_1}$ is less than the standardized fourth moment β_2. Letting $y = (x - \theta)/\lambda$, the probability density function for the unbounded distributions is

$$f(x) = \frac{\partial}{\lambda\sqrt{2\pi}} \frac{1}{\sqrt{1 + y^2}} e^{-\frac{1}{2}\left(\gamma + \delta \sin h^{-1}(y)\right)^2} \qquad -\infty < x < +\infty \quad (5.19)$$

2. *Bounded distributions S_B*: These distributions are used for situations in which the standardized third moment $\sqrt{\beta_1}$ is greater than the standardized fourth moment β_2. The probability density function for the bounded distributions is

$$f(x) = \frac{\partial}{\lambda\sqrt{2\pi}\,y(1-y)}\,e^{-\frac{1}{2}\left(\gamma+\delta\ln\left(\frac{y}{1-y}\right)\right)^2} \quad \theta < x < \theta + \lambda \quad (5.20)$$

3. *Lognormal distributions S_L*: These distributions are used for situations in which the standardized third moment $\sqrt{\beta_1}$ is equal to the standardized fourth moment β_2. The probability density function for the lognormal family of distributions is

$$f(x) = \frac{\partial}{\lambda\sqrt{2\pi}}\,\frac{1}{y}\,e^{-\frac{1}{2}(\gamma+\delta\ln(y))^2} \quad \theta < x < +\infty \quad (5.21)$$

Several methods for fitting a Johnson distribution to a set of data have been proposed. Fitting involves two steps: (1) selecting the appropriate type of distribution from the three shown here and (2) estimating the parameters. Slifker and Shapiro (1980) suggested a method that uses four percentiles of the data to accomplish both steps. A value of the standard normal random variable $0 < Z < 1$ is first selected. The inverse normal cdf is then evaluated at $-3Z$, $-Z$, Z, and $3Z$, which gives four percentages. The percentiles of the data are then calculated at those percentages. Based on the relative distance between the inner and outer percentiles, a transformation function is selected and the parameters estimated. This method is said to be more reliable than matching moments, which tends to be quite variable and greatly influenced by outliers.

Since the Johnson distributions are constructed by transforming the variable X to a standard normal distribution, capability indices may be estimated from the transformed data using the formulas presented in Chapter 4. However, confidence limits for the indices can only be constructed using bootstrapping, since the methodology for selecting and fitting a Johnson curve is difficult to study analytically.

Example 5.7 Transformation using Johnson Curves

Figure 5.8 shows the result of selecting and estimating a Johnson curve for the medical device diameters. The selected distribution is part of the unbounded S_U family. Along the top of the graph are estimates of γ, θ, δ, and λ. As expected, the fitted distribution is positively skewed. Table 5.11 shows confidence limits for several important capability indices obtained using bootstrapping with 50,000 subsamples.

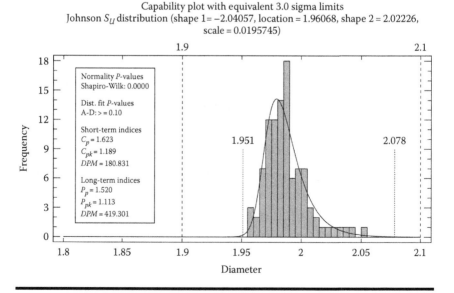

Figure 5.8   Johnson distribution fitted to medical device diameters.

**Table 5.11   Capability Indices for Medical Device Diameters
Estimated Using Fitted Johnson Distribution**

Capability Indices for Diameter		
	Short-Term Capability	*Long-Term Performance*
Sigma (after normalization)	0.936381	1.0
C_p/P_p	1.62332	1.52005
Z_{upper}	3.56664	3.33974
Z_{min}	3.56664	3.33974
C_{pk}/P_{pk}	1.18888	1.11325
DPM	180.831	419.301
Sigma Quality Level	5.07	4.84

Note: Based on 6.0 sigma limits in the normalized metric. Short-term sigma estimated from average moving range. The Sigma Quality Level includes a 1.5 sigma shift in the mean.

95.0% Confidence Bounds—Bootstrap Method (50,000 Subsamples)

Index	Lower Quality Bound
P_p	1.05563
Z_{upper}	2.69506
Z_{min}	2.69348
P_{pk}	0.897825
DPM[a]	3776.34
Sigma Quality Level	4.19348

[a] Lower quality bound corresponds to upper confidence bound for this index.

5.5 Comparison of Methods

The various approaches used in this chapter for dealing with nonnormal data give somewhat different results for the estimated proportion of nonconforming items, as summarized in Table 5.12. The difference in the results is due primarily to differences in the estimated upper tail of the fitted probability distributions. The power transformation approach results in the heaviest upper tail and consequently the highest DPM.

One of the most interesting results is the magnitude of the confidence bounds for P_{pk} and *DPM* when using the Johnson curves compared to the other procedures. The estimated P_{pk} for the Johnson curves is considerably larger than that obtained when using the power transformations, but the lower bound is the same. It may well be that the step involved in identifying the type of Johnson curve adds considerable variability to the estimated capability indices.

In truth, the differences between the three methods for dealing with nonnormal data are not large given the relatively

Table 5.12   Comparison of Results Obtained from Different Methods of Estimating Medical Device Diameter Capability

Method	*Estimated DPM*	*Upper 95% Confidence Bound for DPM*	*Estimated P_{pk}*	*Lower 95% Confidence Bound for P_{pk}*
Power transformation	982	3486	1.03	0.90
Largest extreme value distribution	164	455	1.20	1.11
Johnson curves	419	3776	1.11	0.90
Normal distribution	0.6	9	1.62	1.43

small sample of 100 items. The task of estimating noncon-formance probabilities in the range of several hundred per million from only 100 data values is not an easy one. It would normally be preferable to have at least 300–500 observations in order to choose the best model. Once a model has been established for a particular process variable, smaller samples can then be used to estimate process capability on a regular basis.

References

Box, G.E.P. and Cox, D.R. (1964), An analysis of transforma-tions, *Journal of the Royal Statistical Society, Series B*, **26**, 211–252.

D'Agostino, R.B. and Stephens, M.A. (1986), *Goodness-of-Fit Techniques*, New York: Marcel-Dekker.

Shapiro, S.S. and Wilk, M.B. (1965), An analysis of variance test for normality (complete samples), *Biometrika*, **52**, 591–611.

Slifker, J. and Shapiro, S. (1980), The Johnson system: Selection and parameter estimation, *Technometrics*, **22**, 239–247.

Bibliography

Clements, J.A. (1989), Process capability calculations for non-normal distributions, *Quality Progress*, **22**, 95–100.

Draper, J.R. and Cox, D.R. (1969), On distributions and their trans-formations to normality, *Journal of the Royal Statistical Society, Series B*, **31**, 472–476.

Evans, M., Hastings, N., and Peacock, J.B. (2000), *Statistical Distributions*, 3rd edn., New York: John Wiley & Sons.

George, F. and Ramachandran, K.M. (2011), Estimation of param-eters of Johnson's system of distributions, *Journal of Modern Applied Statistical Methods*, **10**, 2 Article 9.

Hill, I.D., Hill, R., and Holder, R.L. (1976), Algorithm AS 99: Fitting Johnson curves by moments, *Journal of the Royal Statistical Society, Series C (Applied Statistics)*, **25**, 180–189.

Johnson, N.L. (1949), Systems of frequency curves generated by methods of translation, *Biometrika*, **36**, 149–176.

Johnson, N.L., Kotz, S., and Balakrishnan, N. (1994), *Continuous Univariate Distributions*, Vol. 1, 2nd edn., New York: John Wiley & Sons.

Johnson, N.L., Kotz, S., and Balakrishnan, N. (1995), *Continuous Univariate Distributions*, Vol. 2, 2nd edn., New York: John Wiley & Sons.

Sleeper, A. (2007), *Six Sigma Distribution Modeling*, New York: McGraw Hill.

Tukey, J.W. (1957), On the comparative anatomy of transformations, *Annals of Mathematical Statistics*, **28**, 602–632.

Chapter 6

Statistical Tolerance Limits

An increasingly popular method for demonstrating that a process is capable of satisfying established requirements or specifications involves the construction of statistical tolerance limits. Statistical tolerance limits use the information contained in n observations, randomly sampled from a population, to make a statement about a given proportion of that population at a stated level of confidence. For example, $n = 100$ medical devices might be selected from a manufacturing process that produces many such items and the diameter of each of the 100 items measured. A statistical tolerance interval could be constructed from those measurements that would indicate with 95% confidence the range within which the diameter of 99% of all manufactured devices would fall. If this interval is completely within the specification limits, the analyst could state with 95% confidence that at least 99% of all devices would satisfy the specifications.

There are two main types of statistical tolerance limits:

1. *Two-sided intervals* that consist of both a lower limit and an upper limit
2. *One-sided bounds* that consist of one limit only

Two-sided tolerance intervals are useful when the analyst needs to demonstrate that a product falls between two specification limits. One-sided bounds are useful when the analyst needs to demonstrate that a product is either above or below a single specification limit.

Example 6.1 Analysis of Medical Device Diameters

Consider the medical device data introduced in Chapter 1. The diameter of the devices is supposed to fall within the range 2.0 ± 0.1 mm. Suppose that a process engineer wishes to demonstrate with 95% confidence that at least 99% of all devices in the population from which the sample was taken fall within those specification limits.

There are two ways to approach this problem:

1. A distribution could be fit to the data and *parametric* statistical tolerance limits calculated. This would necessitate selecting a particular distributional form, such as a normal distribution (perhaps after transforming the observations) or some other continuous distribution.
2. *Nonparametric* statistical tolerance limits could be calculated that do not assume any specific distribution.

Parametric intervals are usually tighter than the nonparametric intervals since they make assumptions about the shape of the population. However, nonparametric intervals will be correct whether or not those assumptions hold.

6.1 Tolerance Limits for Normal Distributions

If the measurements to be analyzed follow a normal distribution, statistical tolerance limits may be calculated from the sample mean $\bar{x}$ and the sample standard deviation s. Suppose that the engineer wishes to make a statement about $100P\%$ of the population at a confidence level equal to $100(1 - \alpha)\%$. A two-sided normal tolerance interval is calculated by

$$\bar{x} \pm Ks \tag{6.1}$$

where the factor K depends upon the sample size n, the level of confidence $100(1 - \alpha)\%$, and the specified proportion of the population P. A one-sided upper bound is calculated from

$$\bar{x} + K_1 s \tag{6.2}$$

and a one-sided lower bound from

$$\bar{x} - K_1 s \tag{6.3}$$

Note that the K_1 factor for the one-sided bound is not the same as the K factor for the two-sided interval. Tables exist from which the K factors may be obtained for specific combinations of n, P, and α, see for example Montgomery (2013), although most statistical software uses formulas to calculate the K factors.

Example 6.1 (Continued)

For the $n = 100$ medical devices, the sample mean diameter $\bar{x} = 1.98757$ and the sample standard deviation $s = 0.0179749$. To construct a 95% normal tolerance interval for 99% of the population from which the sample was taken, a two-sided "95–99" interval is calculated using $K = 2.93584$:

$$1.98757 \pm 2.93584 * 0.0179749 = (1.93480, 2.04034) \tag{6.4}$$

This results in an interval ranging from approximately 1.935 to 2.040. Figure 6.1 shows the calculated tolerance limits superimposed on a normal distribution with the same mean and standard deviation as the data. Notice that the tolerance limits fit comfortably within the specifications.

Alternatively, if the specification consisted of a single upper specification limit, a 95% upper tolerance bound for 99% of the population could be constructed using $K = 2.68396$:

$$1.98757 + 2.68396 * 0.0179749 = 2.03581 \tag{6.5}$$

The tolerance bound implies that the engineer may be 95% confident that at least 99% of the medical devices in the population are at or below 2.036. As may be seen in Figure 6.2, this bound is closer to the mean of the population than is the upper limit of the two-sided tolerance interval.

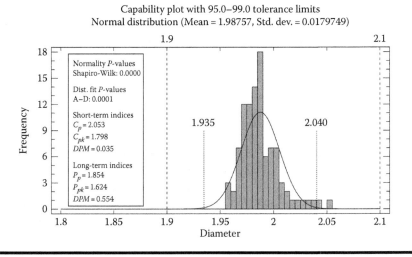

Figure 6.1   95–99 normal tolerance limits for medical device diameters.

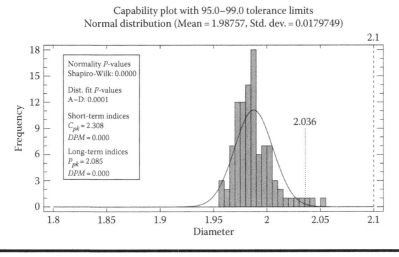

Figure 6.2 95–99 upper tolerance bound for medical device diameters.

It is interesting to note that the central 99% of the fitted distribution of the medical device diameters ranges between 1.94127 and 2.03387 (the 0.5% quantile and the 99.5% quantile), as shown by the shaded area in Figure 6.3. This is tighter than the 95–99 tolerance interval, since the shaded

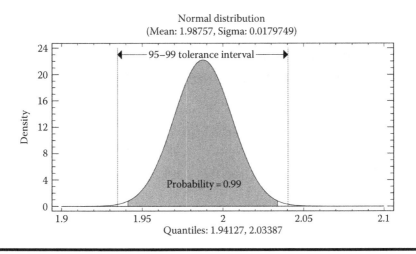

Figure 6.3 Comparison of 95–99 tolerance limits with 0.5% and 99.5% quantiles.

area does not allow for estimation error. The 95–99 statistical tolerance interval allows for potential error in estimating the two quantiles and thus covers more than 99% of the fitted distribution.

6.2 Tolerance Limits for Nonnormal Distributions

The statistical tolerance limits calculated in Section 6.1 are only correct if the data come from a normal distribution. Section 5.1 described the Shapiro-Wilk test, which may be used to determine whether or not the data to be analyzed could reasonably have come from a normal distribution.

   If the assumption of normality does not hold, the process engineer has three choices:

1. Seek a transformation of the data such that after the transformation, the normal distribution is appropriate.
2. Find an alternative distribution that fits the data well and construct statistical tolerance limits based on that distribution.
3. Calculate nonparametric tolerance limits.

6.2.1 Tolerance Limits Based on Power Transformations

Section 5.2 described a method by which a transformation of the data could be sought based on a power p according to

$$X' = (X + \Delta)^p \quad \text{for } p \neq 0 \tag{6.6}$$

and

$$X' = \ln(X + \Delta) \quad \text{for } p = 0 \tag{6.7}$$

This transformation first adds the quantity Δ (called the addend) to each observation and then raises the sum to the power p. Optimal values of p and Δ can often be obtained using the Box-Cox approach. If the transformation is successful, the transformed values should pass the Shapiro-Wilk test for normality.

If a transformation is obtained that yields transformed values that are normally distributed, then statistical tolerance intervals or bounds may be obtained in the transformed metric using the methods of the previous section. The tolerance limits can then be inversely transformed into the original metric, where they will still bound the stated population percentage at the desired confidence level.

Example 6.2 Use of Power Transformations

In Section 5.2, the Box-Cox procedure was used to find an optimal transformation of the medical device diameters. This resulted in values for $p = -3.15$ and $Δ = -1.78688$. The Shapiro-Wilk test was then applied to

$$X' = (X - 1.78688)^{-3.15} \qquad (6.8)$$

which did not reject the hypothesis that the X' values came from a normal distribution. 95% statistical tolerance limits for 99% of the medical device diameters may thus be obtained by

1. Calculating the sample mean and sample standard deviation of the transformed values, $\bar{x} = 164.735$ and $s = 40.6898$
2. Constructing tolerance limits in the transformed metric from

$$164.735 \pm 2.93584 * 40.6898 = (45.2763, 284.194) \quad (6.9)$$

3. Applying the inverse transformation to the values in (6.9)

$$LTL = 284.194^{1/-3.15} + 1.78688 = 1.95325 \qquad (6.10)$$

$$UTL = 45.2763^{1/-3.15} + 1.78688 = 2.08496 \qquad (6.11)$$

Figure 6.4 shows the inverted statistical tolerance limits together with the specification limits and the implied distribution of the medical device diameters. Note that the implied distribution, which is the fitted normal distribution subjected to the inverse transformation, is skewed to the right in the same manner as the data. Compared with the statistical tolerance limits calculated under the assumption that the original measurements come from a normal distribution, the tolerance limits based on the transformation approach are both shifted to the right.

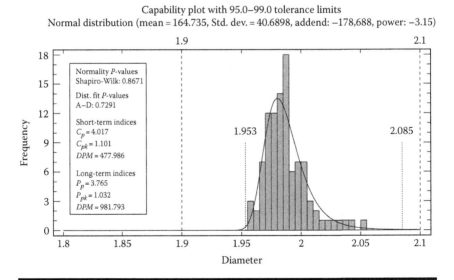

Capability plot with 95.0–99.0 tolerance limits
Normal distribution (mean = 164.735, Std. dev. = 40.6898, addend: −178,688, power: −3.15)

Figure 6.4   95–99 tolerance limits for medical device diameters based on normal distribution after transformation.

6.2.2 Tolerance Limits Based on Alternative Distributions

If a transformation cannot be found that makes the transformed data normally distributed, an alternative distribution can be sought. Methods exist for obtaining statistical tolerance limits based on various other distributions, including the Cauchy, exponential, 2-parameter exponential, gamma, Laplace, largest extreme value, lognormal, Pareto, smallest extreme value, and Weibull distributions. If one of these distributions fits the data, then it may be used to calculate the intervals. Patel (1986) gives a useful review of the methodologies.

As an example, suppose the data values are well modeled by a largest extreme value distribution. Let $\hat{\gamma}$ be the estimated mode and $\hat{\beta}$ be the estimated scale parameter when that distribution is fit to a sample of n observations. Single lower and upper one-sided tolerance limits are given by

$$LTL = \hat{\gamma} + \frac{\hat{\beta} \, t^*_{n-1,\alpha} \left(-\sqrt{n} \ln \left(-\ln \left(1 - P \right) \right) \right)}{\sqrt{n-1}} \qquad (6.12)$$

$$UTL = \hat{\gamma} + \frac{\hat{\beta} \, t^*_{n-1,1-\alpha} \left(-\sqrt{n} \ln \left(-\ln \left(P \right) \right) \right)}{\sqrt{n-1}} \qquad (6.13)$$

where $t^*_{n-1,\alpha}(\Delta)$ is the αth quantile of a noncentral t distribution with $n-1$ degrees of freedom and noncentrality parameter Δ. Two-sided tolerance limits are calculated by replacing α with $\alpha/2$ in these formulas and by replacing P with $(P + 1)/2$.

Example 6.3 Tolerance Limits Based on Largest Extreme Value Distribution

In Section 5.3, it was found that the medical device diameters were well modeled by a largest extreme value distribution with $\hat{\gamma} = 1.97962$ and $\hat{\beta} = 0.0138082$. Setting $n = 100$, $\alpha = 0.025$, and $P = 0.995$ in Equations 6.12 and 6.13 gives the values $LTL = 1.952$ and $UTL = 2.065$. The tolerance limits and fitted distribution are shown in Figure 6.5.

Table 6.1 compares the estimated tolerance limits obtained by the three methodologies described so far in this chapter. The power transformation and fitted largest extreme value distribution give very similar values for the lower tolerance limit, both well above that obtained using the normal distribution. This is clearly due to the much shorter lower tail. There is a larger difference between the estimated upper tolerance limits. The power transformation approach is based on an implied distribution that is heavier in the upper tail than the largest extreme value distribution, resulting in a larger upper tolerance limit. This is consistent with the smaller value of P_{pk} found when using the power transformation in Chapter 5.

Note that all three approaches give intervals that are well inside of the specification limits.

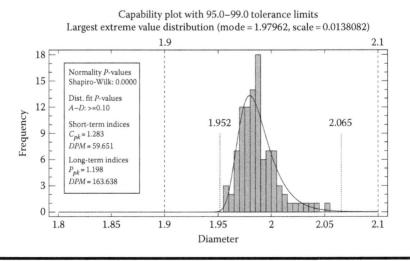

Capability plot with 95.0–99.0 tolerance limits
Largest extreme value distribution (mode = 1.97962, scale = 0.0138082)

Figure 6.5 Statistical tolerance limits for medical device diameters based on fitted largest extreme value distribution.

Table 6.1   95–99 Statistical Tolerance Limits for the Medical Device Diameters Using Three Different Methods

Method	Lower Tolerance Limit	Upper Tolerance Limit
Fitted normal distribution	1.935	2.040
Optimized power transformation	1.953	2.085
Fitted largest extreme value distribution	1.952	2.065

6.3 Nonparametric Statistical Tolerance Limits

It is also possible to construct statistical tolerance limits without making any assumption about the shape of the distribution from which the observations were obtained. Given n observations randomly sampled from some unknown population, the data are first sorted to obtain the order statistics $X_{(1)}, X_{(2)}, \ldots, X_{(n)}$. A tolerance interval is then constructed from

$$\left[X_{(d)}, X_{(n-d+1)} \right] \tag{6.14}$$

where d is a number such as 1, 2, or 3, called the "depth". If $d = 1$, the tolerance interval is created from the smallest and largest observations. If $d = 2$, it is created from the second smallest and second largest observations, and so on.

For any given value of d, the internal in (6.14) contains at least $100P\%$ of the population with $100(1 - \alpha)\%$ confidence where

$$P = (q - 1)/(q + 1) \tag{6.15}$$

$$q = \frac{4(n - d + .5)}{X^2_{\alpha,4d}} \tag{6.16}$$

The analyst can choose to set either P or α, but not both. For example, if $P = 0.99$ then α is determined by the equations. Likewise, if $\alpha = 0.05$ then P is determined by the equations.

Example 6.4 Nonparametric Tolerance Limits

Suppose that nonparametric tolerance limits are desired for the medical device data. For that data, $n = 100$. Setting the depth $d = 1$ and letting $\alpha = 0.05$, the interval formed by the largest and smallest observations [1.956, 2.053] turns out to be a 95% tolerance interval for 95.3433% of the population, as shown in Figure 6.6. Setting $P = 0.99$ instead, that same interval turns out to be a 26.4% tolerance interval for 99% of the population, as shown in Figure 6.7. The only way to obtain a 95–99 nonparametric tolerance interval would be to increase the sample size n substantially. Sample size determination for statistical tolerance limits is discussed in Chapter 8.

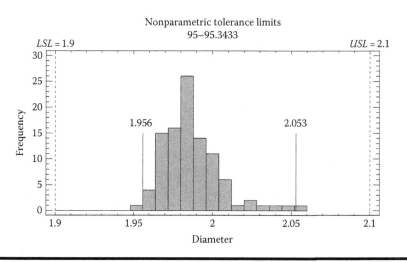

Figure 6.6 Nonparametric tolerance interval for medical device diameters formed by setting $\alpha = 0.05$.

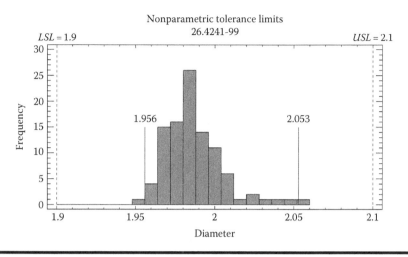

Figure 6.7   Nonparametric tolerance interval for medical device diameters formed by setting *P* = 0.99.

References

Montgomery, D.C. (2013), *Introduction to Statistical Quality Control*, Seventh edition, Hoboken, NJ: John Wiley& Sons.
Patel, J.K. (1986), Tolerance limits: A review, *Communication in Statistics: Theory and Methods*, **15**, 2719–2762.

Bibliography

Hahn, G.J., Meeker, W.Q., and Escobar, L.A. (2017), *Statistical Intervals: A Guide for Practitioners and Researchers*, 2nd edn., New York: Wiley.
Krishnamoorthy, K. and Mathew, T. (2009), *Statistical Tolerance Regions: Theory, Applications, and Computation*, Hoboken, NJ: John Wiley & Sons.
Statgraphics Technologies, Inc. (2017), Statistical tolerance limits (observations), PDF documentation for Statgraphics Centurion 18, the Plains, VA.

Chapter 7

Multivariate Capability Analysis

All of the material in the previous chapters deals with the estimation of process capability for a single variable. For many processes, acceptable performance requires that multiple variables be within spec. If the variables are independent and the occurrence of nonconforming items is very low, the joint probability of an item being out of spec with respect to one or more of the variables will be very close to the sum of the individual nonconformance probabilities for each of the variables. However, if the variables are strongly correlated, adding the separate probabilities may not give a good estimate of the combined nonconformance probability. Also, if the specifications contain requirements for the joint behavior of two or more variables, those variables must be considered together.

This chapter deals with the simultaneous estimation of process capability with respect to more than one variable. Given m variables, θ will represent the probability that a randomly selected item from a process is out of spec with respect to one or more of the variables. Estimation of θ requires constructing a multivariate distribution to describe the joint behavior of those variables.

Example 7.1 Bivariate Data Visualization

Consider a process for manufacturing a medical device. As in earlier chapters, one of the variables of concern in that process is the diameter of the devices X_1, which must be between 1.9 and 2.1. A second variable of interest is the *strength* of the devices X_2, which must not be less than 200 psi. To estimate the probability of not conforming to the specifications with respect to either or both of the variables, a sample of $n = 200$ items has been obtained.

Figure 7.1 plots the measured diameter and strength of the 200 devices. It shows that there is a strong positive correlation between the 2 variables. The acceptable region in the space of diameter and strength is an unbounded rectangle with sides at $X_1 = 1.9$ and $X_1 = 2.1$ and a lower bound at $X_2 = 200$. In the discussion that follows, this data will be used to estimate the probability of producing a device that is outside of the acceptable region.

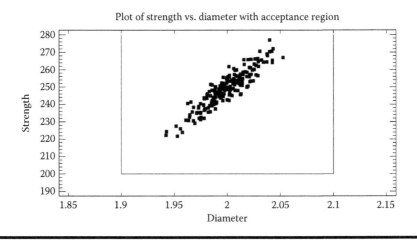

Figure 7.1 Scatterplot of strength and diameter for sample of 200 medical devices.

7.1 Visualizing Bivariate Data

To visualize the joint distribution of 2 variables, it is useful to plot a *bivariate histogram*. A bivariate histogram divides the range of values for each variable into classes and then displays the number of observations that fall within each combination of classes.

Example 7.1 (Continued)

Figure 7.2 shows a bivariate histogram for the medical device data. The range of the variables has been divided into a 20 by 20 grid. The height of the bars is proportional to the number of devices with diameter and strength corresponding to each cell of the grid.

Given data on m variables, a probability density function $f_X(X_1, X_2, \ldots, X_m)$ may be defined from which probabilities may be calculated for joint values of those variables. In particular, the probability that the m variables fall within some region is the multiple integral of the density function over that region. The next section considers the use of a multivariate normal distribution for modeling the density function.

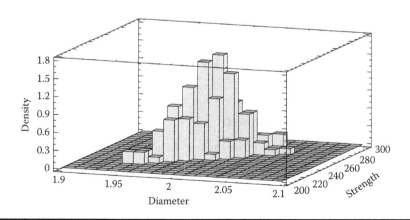

Figure 7.2   Bivariate histogram for the medical device data.

Prior to estimation of a parametric model, it is instructive to display a nonparametric estimate of the density function. A popular method for estimating the density function is to pick multiple locations throughout the range of the data and calculate a weighted count of the nearby observations. A widely used estimate of a bivariate density function at location (X_1, X_2) is given by

$$\hat{f}(X_1, X_2) = \frac{(\det S)^{-1/2}}{b^2 n} \sum_{i=1}^{n} W\left(\frac{1}{b^2}(X_{1,i} - X_1)^T S^{-1}(X_{2,i} - X_2)\right)$$

(7.1)

where
   S is the sample covariance matrix of the 2 variables
   b is the bandwidth
   $W(u)$ is a weighting function defined by

$$W(u) = \frac{1}{2\pi} \exp(-u/2)$$
(7.2)

The bandwidth b determines how quickly the weighting function decays, with larger values giving more weight to points far from the position at which the density is being estimated. A bandwidth of 0.5 is not unreasonable for a small sample but may not give as much detail as a smaller bandwidth in larger samples.

Example 7.1 (Continued)

Figure 7.3 shows a nonparametric estimate of the bivariate density function for diameter and strength obtained from the medical device data, using a bandwidth of $b = 0.3$. The estimated density function has a well-defined peak and is elongated in a direction corresponding to positive correlation between the variables.

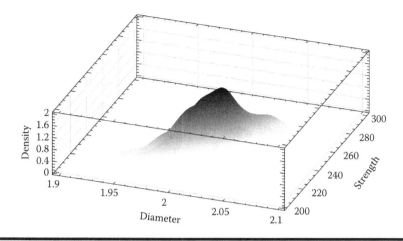

Figure 7.3   Nonparametric estimate of density function for medical device diameter and strength.

7.2 Multivariate Normal Distribution

The most widely used model for multivariate data is the multivariate normal distribution. Given a vector of random variables $\boldsymbol{X} = (X_1, X_2, ..., X_m)$, the multivariate normal probability density function is

$$f_X\left(X_1, X_2, ..., X_m\right) = \frac{1}{\sqrt{(2\pi)^m |\Sigma|}} \exp\left(-\frac{1}{2}(X - \mu)^T \Sigma^{-1}(X - \mu)\right)$$

$$(7.3)$$

where
  μ is a vector of means
  Σ is an m by m covariance matrix

For the bivariate case, the vector of means is

$$\mu = \begin{pmatrix} \mu_1 \\ \mu_2 \end{pmatrix}$$

$$(7.4)$$

and the covariance matrix is

$$\Sigma = \begin{pmatrix} \sigma_1^2 & \rho\sigma_1\sigma_2 \\ \rho\sigma_1\sigma_2 & \sigma_2^2 \end{pmatrix} \tag{7.5}$$

where

σ_1 is the standard deviation of variable 1
σ_2 is the standard deviation of variable 2
ρ is the correlation between the two variables

The marginal distribution of each individual variable X_j is a univariate normal distribution with mean μ_j and standard deviation σ_j.

Fitting a multivariate normal distribution to n multivariate observations requires estimating m means, m variances, and $m(m-1)$ covariances. Letting $X_{i,j}$ represent the value of the ith observation for variable j, unbiased estimates of the means are given by the sample means:

$$\hat{\mu}_j = \bar{x}_j = \frac{\sum_{i=1}^{n} X_{i,j}}{n} \tag{7.6}$$

Unbiased estimates of the variances are given by the sample variances:

$$\hat{\sigma}_j^2 = s_j^2 = \frac{\sum_{i=1}^{n} \left(X_{i,j} - \bar{X}_j \right)^2}{n-1} \tag{7.7}$$

An unbiased estimate of the covariance between variables j and k is given by the sample covariance:

$$\hat{\sigma}_{j,k} = \frac{\sum_{i=1}^{n} \left(X_{i,j} - \bar{X}_j \right)\left(X_{i,k} - \bar{X}_k \right)}{n-1} \tag{7.8}$$

Example 7.2 Fitting a Multivariate Normal Distribution

Figure 7.4 shows a multivariate normal distribution fit to the $n = 200$ observations of diameter and strength for the medical devices. It has a well-defined peak at the centroid $(\hat{\mu}_1, \hat{\mu}_2) = (1.99958, 249.300)$. The standard deviations are $(\hat{\sigma}, \hat{\sigma}_2) = (0.0208047, 10.4658)$ and the sample covariance $\hat{\sigma}_{1,2} = 0.199052$.

It is also instructive to calculate the correlation between the 2 variables defined by

$$\hat{\rho}_{j,k} = \frac{\hat{\sigma}_{j,k}}{\hat{\sigma}_i \hat{\sigma}_j} \qquad (7.9)$$

By definition, the correlation ranges between -1 and $+1$, with values close to -1 indicating a strong negative correlation between the variables and values close to $+1$ indicating a strong positive correlation. For the diameter and strength of the medical device data, $\hat{\rho}_{1,2} = 0.9142$. This strong positive correlation accounts for the elongation of the major axis of the ellipses in Figure 7.5, which contain 25%, 50%, 75%, 90%, 95%, and 99% of the fitted distribution.

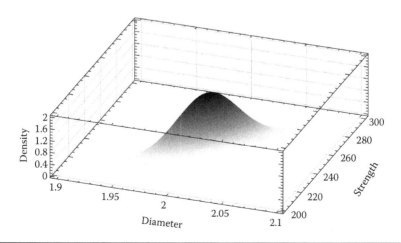

Figure 7.4   Bivariate normal density function fitted to diameter and strength of medical devices.

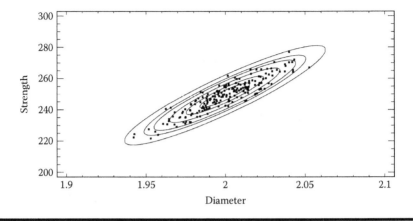

Figure 7.5 Contour plot of bivariate normal density function for medical device data.

7.3 Multivariate Tests for Normality

Before calculating statistics that rely on the assumption that data come from a multivariate normal distribution, it is a good idea to test the assumption of multivariate normality. In particular, given a random sample of n observations in m dimensions, the following hypotheses should be tested:

> *Null hypothesis*: Data are a random sample from a multivariate normal distribution.
> *Alternative hypothesis*: Data are not a random sample from a multivariate normal distribution.

One of the best tests for multivariate normality is due to Royston (1983). He derived a test statistic H that combines the Shapiro-Wilk W statistics calculated for each of the m variables separately. H is referred to a chi-square distribution with degrees of freedom that depend on the correlations among the original variables.

Example 7.3 Tests for Multivariate Normality

Table 7.1 shows the results obtained when Roysten's test is run on the medical device data. The table includes the Shapiro-Wilk W test for each of the variables and Roysten's H statistic. Since the P-value of Roysten's H is well above 0.05, there is no evidence that a bivariate normal distribution would not be a good model for the data.

A second way to test the hypothesis of multivariate normality is by creating a chi-square plot. Let $\textbf{X}_i = (X_{i,1}, X_{i,2}, ..., x_{i,m})$ be a column vector with the observed values of each variable

Table 7.1   Results of Roysten's Test of Multivariate Normality for Medical Device Diameter and Strength

Multivariate Normality Test		
Number of observations = 200		
	Mean	*Standard Deviation*
Diameter	1.99958	0.0208047
Strength	249.3	10.4658
Sample Correlations		
	Diameter	*Strength*
Diameter	1.0	0.914186
Strength	0.914186	1.0
Normality Tests		
Test	*Statistic*	*P-Value*
Shapiro-Wilk W—diameter	0.997	0.9378
Shapiro-Wilk W—strength	0.993	0.5058
Roston's H	0.315	0.7178

for the ith observation. Let $\bar{X}$ be a vector containing the m sample means and let S represent the m by m sample covariance matrix. Then if the data come from a multivariate normal distribution, the squared generalized distances

$$d_i^2 = \left(X_i - \bar{X} \right)^T S^{-1} \left(X_i - \bar{X} \right), \quad i = 1, 2, \ldots, n \qquad (7.10)$$

should behave like a random sample from a chi-square distribution with m degrees of freedom. A simple approach to testing for multivariate normality is to construct the squared distances, fit a chi-square distribution to them, and use a standard univariate goodness-of-fit test to determine whether or not they could reasonably have come from that distribution.

Example 7.3 (Continued)

Figure 7.6 shows a chi-square plot constructed for the bivariate medical device data. The CDF of the fitted chi-square distribution is plotted, together with 95% Kolmogorov-Smirnov limits. The empirical CDF of the generalized distances is shown by the point symbols, calculated from the 200 observations. The points fall well inside the K-S limits, as they should if they come from the hypothesized chi-square distribution.

Also included on the graph is the result of an Anderson-Darling test comparing the squared distances to a chi-square distribution with 2 degrees of freedom. The P-value is well above 0.05, indicating that the distances are well modeled by the chi-square distribution. This result leads in turn to the conclusion that the original data could well have come from a bivariate normal distribution.

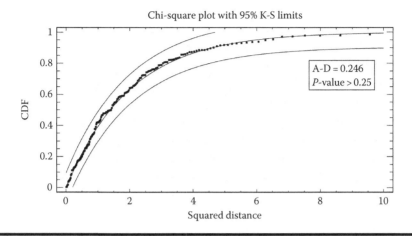

Figure 7.6 Chi-square plot of squared generalized distances for bivariate medical device data.

7.4 Multivariate Capability Indices

Estimating process capability based on the multivariate normal distribution requires estimating θ, the probability of being out of spec with respect to one or more of the variables. Given estimates of the variable means and covariance matrix as calculated earlier, this probability may be estimated by setting $\hat{\theta}$ equal to 1 minus the integral of the fitted multivariate density function $\hat{f}_X(X_1, X_2,\ldots, X_m)$ over the acceptable region as defined by the specification limits. This integration must be done numerically, which is straightforward but may be time-consuming in high dimensions. For variables that are not highly correlated, the result is likely to be quite close to the sum of the individual nonconformance probabilities. For highly correlated variables, however, it may be much different.

Example 7.4 Multivariate Capability Indices

It will be recalled that the specifications for the medical devices call for the diameter to be between 1.9 and 2.1 and the strength to be greater than or equal to 200. Table 7.2 shows the results of fitting a multivariate normal distribution to the $n = 200$ sample devices. The estimated defects per million based on a bivariate analysis of diameter and strength is approximately 2.22. Note that this number is only about 80% of the sum of the estimated DPMs for the two variables when analyzed separately. This is due to a strong positive correlation between the variables ($\hat{\rho} = 0.914$), which means that there is a significant probability of defects occurring with respect to both variables simultaneously.

Table 7.2   Estimated Defects per Million Based on Fitted Multivariate Normal Distribution for Medical Device Data

Multivariate Capability Analysis					
Number of complete cases: 200					
Variable	Sample Mean	Sample Std. Dev.	LSL	Nominal	USL
Diameter	1.99958	0.0208047	1.9	2.0	2.1
Strength	249.3	10.4658	200.0		

Variable	Observed Beyond Spec	Estimated Beyond Spec	Estimated DPM
Diameter	0.0%	0.000154463%	1.54463
Strength	0.0%	0.000123616%	1.23616
Joint	0.0%	0.000221754%	2.21754

A good way to illustrate the result of a multivariate capability analysis based on two variables is to draw an ellipse in the space of those variables that contains 99.73% of the bivariate normal distribution. This is the same percentage as is bound by $\mu \pm 3\sigma$ when analyzing a single variable. If the entire region bound by the ellipse is within the specification region, the capability of the bivariate process to meet the specification limits will be at least as good as that of a univariate process with a capability index $P_{pk} = 1.0$. If desired, the percentage bound by the ellipse may be increased so that it corresponds to a larger capability index.

The equation for determining the ellipse that contains $100(1 - \alpha)\%$ of a fitted bivariate normal distribution is given by

$$\chi^2_{2,\alpha} = \left(X - \hat{\mu} \right)^T S^{-1} \left(X - \hat{\mu} \right) \tag{7.11}$$

where $\chi^2_{2,\alpha}$ is the value of a chi-square distribution with 2 degrees of freedom that is exceeded with probability equal to α.

Example 7.4 (Continued)

Figure 7.7 shows a 99.73% capability ellipse, drawn using the fitted bivariate normal distribution for the medical device data. The rectangular region represents the acceptable region as defined by the specification limits. The ellipse fits comfortably within the boundaries of that region.

Several multivariate capability indices may be constructed in a way that allows them to be treated in the same manner as for a single variable. The first is an equivalent Z index.

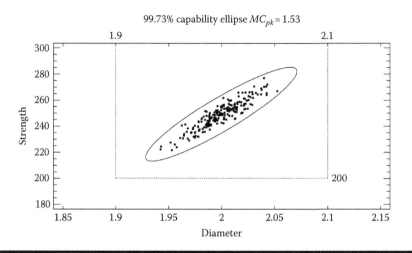

Figure 7.7 Capability ellipse containing 99.73% of the fitted bivari-ate normal distribution for medical device data.

Having estimated the proportion of nonconforming items $\hat{\theta}$, an equivalent Z index may be calculated from

$$Z = \Phi^{-1}\left(1 - \hat{\theta}\right) \qquad (7.12)$$

where Φ^{-1} is the inverse standard normal cumulative distribu-tion function. Simply stated, Z is the value of a standard nor-mal random variable, which is exceeded with probability $\hat{\theta}$. Interpretation of Z is similar to the univariate case, with a typical target value being $Z \geq 4$.

The sigma quality level may also be calculated in the usual manner from

$$SQL = Z + 1.5 \qquad (7.13)$$

As in the univariate case, Six Sigma practitioners would define a process as producing a product or service with "world class quality" if $SQL \geq 6$.

A multivariate version of C_{pk} may also be calculated from

$$MC_{pk} = \frac{Z}{3} \qquad (7.14)$$

Table 7.3 Estimated Multivariate Capability Indices for the Medical Device Diameters and Strength

Capability Indices	
Index	*Estimate*
MC_{pk}	1.53
DPM	2.21754
Z	4.58989
SQL	6.08989

Note: Based on 6.0 sigma limits. The Sigma Quality Level includes a 1.5 sigma shift in the means.

This index retains the same relationship with the estimated proportion of nonconforming items as does the univariate P_{pk}. Consequently, MC_{pk} should normally be at least 1.33.

Example 7.4 (Continued)

Table 7.3 displays multivariate capability indices for the medical device data based on both diameter and strength. The process is estimated to be operating at slightly over a 6 sigma level.

7.5 Confidence Intervals

To estimate confidence intervals for the proportion of non-conforming items and for capability indices, the only choice available is to use bootstrapping. As will be recalled from Chapter 5, this requires

1. Selecting subsamples of *n* observations from the sample data at random *with replacement*

2. Fitting the multivariate normal distribution and calculating the capability indices
3. Repeating steps 1 and 2 many times in order to create a sampling distribution for the indices
4. Calculating the desired percentiles using the bootstrapped sampling distributions

Example 7.4 (Continued)

Table 7.4 displays lower 95% confidence bounds for the multivariate capability indices based on the fitted bivariate normal distribution for diameter and strength. They are based on the creation of 5000 random subsamples of the original 200 observations. Based on the intervals in that table, it may be stated with 95% confidence that the proportion of devices that do not meet the specifications on one or both variables is no greater than 12.9 per million.

Table 7.4   Confidence Bounds for Multivariate Capability Indices for Medical Device Data Calculated Using Bootstrapping

95.0% Confidence Bounds: Bootstrap Method (5000 Subsamples)	
	Lower Limit
MC_{pk}	1.40278
DPM[a]	12.8636
Z	4.20833
SQL	5.70833

[a] Lower quality bound corresponds to upper confidence bound for this index.

7.6 Multivariate Normal Statistical Tolerance Limits

Rather than calculating multivariate capability indices, it is possible instead to create statistical tolerance limits based on the multivariate normal distribution and compare those tolerance limits to the joint specifications for the *m* variables. This is particularly useful when the specifications for the variables are more complicated than just upper and lower bounds for each of the variables. For example, there could be a specification based on the ratio of 2 variables or on some linear or nonlinear combination of more than one variable. Given statistical tolerance limits, it is possible to determine whether or not the entire joint tolerance region meets the specifications.

Two primary approaches to the construction of tolerance limits for multivariate data will be considered:

1. Construction of a joint tolerance region for the *m* variables. This is a region in *m* dimensions that contains a proportion *P* of the joint distribution of the variables with $100(1 - \alpha)\%$ confidence. Construction of the region will involve using a Monte Carlo simulation to obtain a critical parameter.
2. Construction of simultaneous tolerance limits for each of the *m* variables individually, adjusting the confidence level using a Bonferroni approach. This is a somewhat conservative approach but easy to implement and interpret.

7.6.1 Multivariate Tolerance Regions

Consider a random sample of *n* observations from an *m*-dimensional multivariate normal distribution. Let $\bar{X}$ be an *m* by 1 column vector containing the sample means of each variable and let S be the m by m sample covariance matrix. A joint tolerance region for the m variables is given by

$$\left(X - \bar{X}\right)^T S^{-1} \left(X - \bar{X}\right) \leq c \qquad (7.15)$$

where c is a constant that depends on m, n, P and α. For $m = 2$, the region is an ellipse. For $m > 2$, the region is an ellipsoid or a hyperellipse.

Unfortunately, there is no exact method for obtaining c, nor are the available approximations adequate for all combinations of m, n, P, and α. Krishnamoorthy and Mathew (2009) suggest that the best way to obtain a value for c is to use Monte Carlo simulation. In particular, their Algorithm 9.2 uses an approach that involves the generation of random variables from chi-square and Wishart probability distributions.

Example 7.5 Multivariate Normal Tolerance Region

Returning to the sample data, suppose a 95% statistical tolerance region is desired for the diameter and strength of 99.9% of the medical devices being produced. In the space of the two response variables, the region is an ellipse defined by Equation 7.15 where c is determined from $m = 2$, $n = 200$, $P = 99.9\%$, and $(1 - \alpha) = 95\%$.

Table 7.5 displays the estimated 95% tolerance region for 99.9% of all medical devices in the population. The output indicates that $c = 16.0526$, based on a Monte Carlo simulation using 100,000 subsamples. Figure 7.8 displays the tolerance region graphically as an ellipse. All of the 200 bivariate observations are within the tolerance region. In summary, it may be stated with 95% confidence that at least 99.9% of all medical devices being manufactured fall within the elliptical region. Since the elliptical region is completely within the specification limits, it may also be stated with 95% confidence that at least 99.9% of all the devices will be within the specification limits.

Table 7.5 Multivariate Tolerance Limits for the Diameter and Strength of the Medical Devices

Multivariate Tolerance Limits		
Number of observations = 200		
95% Elliptical Tolerance Region for 99.9% of the Population: Squared distance ≤ 16.0526		
Observations outside elliptical region: 0		
95% Simultaneous Bonferroni Tolerance Limits for 99.9% of the Population		
	Lower Limit	*Upper Limit*
Diameter	1.92346	2.0757
Strength	213.12	
Observations beyond Bonferroni limits: 0		

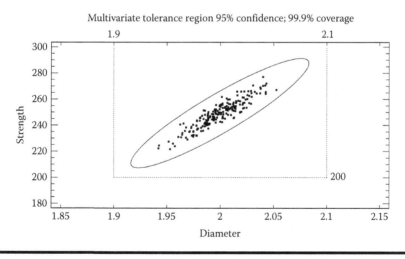

Figure 7.8   95–99.9 tolerance region for the medical devices.

7.6.2 Simultaneous Tolerance Limits

A second way to create tolerance limits for multiple variables is to calculate univariate tolerance limits for each variable separately, adjusting the confidence level of each tolerance limit such that all of the limits will be correct simultaneously with at least the stated level of confidence. Recall that the tolerance limit for variable j is calculated using the sample mean and sample standard deviation of that variable as discussed in Section 6.1:

$$\bar{x}_j \pm Ks_j \qquad (7.16)$$

If limits are calculated for each of the m variables with confidence level set to $100(1 - \alpha/m)\%$, all limits will be correct at least $100(1 - \alpha)\%$ of the time. Although this approach is somewhat conservative (the true confidence level may be higher than that stated), it makes it very simple to compare the tolerance limits to the specifications.

Example 7.5 (Continued)

Figure 7.9 shows a shaded region corresponding to 95% simultaneous tolerance limits for 99.9% of the distribution of diameter and strength for the medical devices, based on separate 97.5% tolerance limits for each of the two variables. As displayed in Table 7.5, the limits are

$$1.92346 \leq diameter \leq 2.0757$$
$$213.12 \leq strength$$

As with the elliptical tolerance region, all 200 sample points are within the tolerance limits and the tolerance limits are completely within the specification limits.

Figure 7.10 shows the tolerance regions calculated using both approaches. Note that the elliptical tolerance region, which accounts for the correlation between diameter and strength, has a smaller area than the Bonferroni limits. It is also interesting to note that some area within the ellipse is not contained within the shaded area, and vice versa. Clearly, the two methods are allocating the 0.1% beyond the tolerance limits to different regions within the space of diameter and strength.

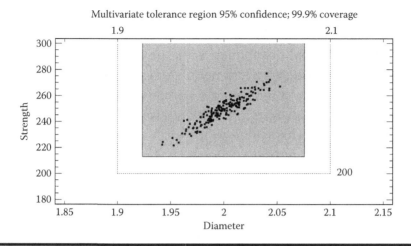

Figure 7.9   Simultaneous tolerance limits for the diameter and strength of the medical devices using the Bonferroni method.

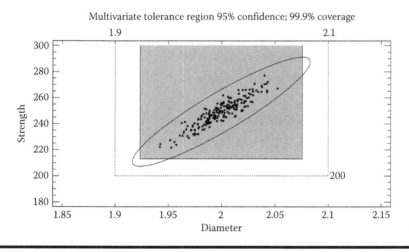

Figure 7.10   Elliptical and Bonferroni tolerance regions for the diameter and strength of the medical devices.

7.7 Analysis of Nonnormal Multivariate Data

If the multivariate data to be analyzed do not come from a normal distribution, transformations may be sought that normalize the data. Often, either a Box-Cox power transformation or a Johnson transformation applied to each variable separately will make the multivariate normal distribution a reasonable model for the data. For example, one widely used alternative to the multivariate normal distribution is the multivariate lognormal distribution. For such a distribution, the logarithms of the variables jointly follow a multivariate normal distribution.

Individually transforming each variable has some drawbacks, however. This is due to the fact that multivariate normality implies three basic properties:

1. The marginal distribution of each variable is normal.
2. The expectation of any subvector X_1 given any other subvector X_2 is linear. For example, in the bivariate case

$$E[X_1|X_2] = \mu_1 + \frac{\sigma_{1,2}(X_2 - \mu_2)}{\sigma_2^2} \qquad (7.17)$$

3. The conditional variance of X_1 given X_2 is constant. This is similar to the assumption of homoscedasticity (constant variance) that applies in the univariate case.

There is no guarantee that individually transforming the variables to achieve property (1) will also result in properties (2) and (3) being satisfied.

Andrews et al. (1971) proposed a multivariate transformation that attempts to find powers that maximize the likelihood of the multivariate normal distribution for the transformed variables when considered simultaneously. It assumes that there is a vector of powers $\lambda = (\lambda_1, \lambda_2, ..., \lambda_m)$ that when applied to the m variables transforms them to a multivariate normal distribution. Maximum likelihood estimates of the powers may be obtained by maximizing the profile likelihood given by

$$-\frac{n \log|S(\lambda)|}{2} + \sum_{j=1}^{m}\left\{(\lambda_j - 1)\sum_{i=1}^{n}\log(X_{i,j})\right\} \qquad (7.18)$$

where $S(\lambda)$ is the estimated covariance matrix of the transformed variables. Studies have shown that maximization of (7.18) results in more efficient estimation of the powers than determining each transformation separately.

Maximization of (7.18) requires a numerical solution, which is easily performed by most statistical software. If necessary, each variable can be shifted to a new origin using a 2-parameter transformation with an addend Δ_j, in which case $X_{i,j}$ is replaced by $X_{i,j} + \Delta_j$ in Equation 7.18.

Example 7.6 Analysis of Bivariate Lognormal Data

Figure 7.11 shows a bivariate histogram for 2 variables that were randomly sampled from a multivariate lognormal distribution. There is a noticeably longer tail toward the upper right corner than toward the lower left corner. Table 7.6 shows the results of applying Roysten's test for multivariate normality. The P-value is very small, rejecting the hypothesis that the data come from a multivariate normal distribution.

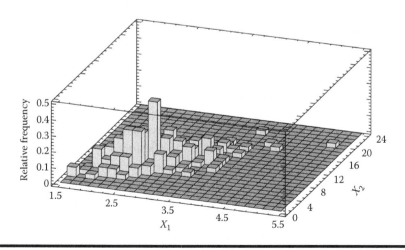

Figure 7.11   Bivariate histogram for simulated multivariate lognormal data.

Table 7.6 Multivariate Test for Normality for Simulated Lognormal Data

Multivariate Normality Test		
	Mean	*Standard Deviation*
X_1	2.78161	0.543724
X_2	8.05924	3.21711
Sample Correlations		
	X_1	X_2
X_1	1.0	0.874526
X_2	0.874526	1.0
Normality Tests		
Test	*Statistic*	*P-Value*
Shapiro-Wilk $W-X_1$	0.970	0.0003
Shapiro-Wilk $W-X_2$	0.939	0.0000
Royston's H	31.212	0.0000

Example 7.6 (Continued)

Applying a power transformation to each variable separately using the Box-Cox procedure gives the estimates $\hat{\lambda}_1 = -0.075$ and $\hat{\lambda}_2 = 0.076$. Table 7.7 shows that determining the powers simultaneously by maximizing the profile likelihood in (7.18) gives the estimates $\hat{\lambda}_1 = -0.080$ and $\hat{\lambda}_2 = 0.073$. The estimated values of both λ_1 and λ_2 are close to 0, which corresponds to a power transformation involving logarithms as discussed in Section 5.2.

Table 7.7 also shows the tests for multivariate normality after applying the simultaneous transformation. Note that the P-value for Roysten's test is now quite large, suggesting that the procedure has produced transformed variables which are adequately modeled by a multivariate normal distribution. Figure 7.12 shows the nonparametric

Table 7.7 Multivariate Test for Normality after Multivariate Box-Cox Transformation

Multivariate Normality Test		
Power transformations: estimated simultaneously		
Variable	*Power*	
X_1	−0.080127	
X_2	0.0733825	
	Mean	*Standard Deviation*
X_1	0.922769	0.0142053
X_2	1.15951	0.0333334
Sample Correlations		
	X_1	X_2
X_1	1.0	−0.880659
X_2	−0.880659	1.0
Normality Tests		
Test	*Statistic*	*P-Value*
Shapiro-Wilk $W-X_1$	0.996	0.9192
Shapiro-Wilk $W-X_2$	0.997	0.9764
Royston's H	0.009	0.9834

estimate of the density of the transformed variables, which looks much like that expected for data from a multivariate normal distribution.

Having found a transformation of the variables that achieves multivariate normality, the acceptable region defined by the specification limits is then transformed in a similar manner. Quality indices and statistical tolerance limits are calculated in the transformed metric.

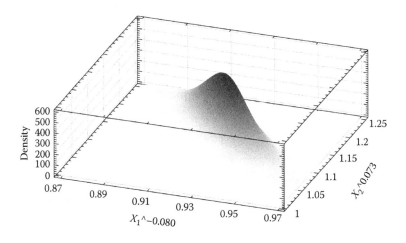

Figure 7.12   Nonparametric estimate of bivariate density for trans-formed variables.

References

Andrews, D.F., Gnanadesikan, R., and Warner, J.L. (1971),
    Transformations of multivariate data, *Biometrics*, **27**, 825–840.
Krishnamoorthy, K. and Mathew, T. (2009), *Statistical Tolerance
    Regions: Theory, Applications, and Computation*, Hoboken, NJ:
    John Wiley & Sons.
Royston, J.P. (1983), Some techniques for assessing multivariate
    normality based on the Shapiro-Wilk W, *Applied Statistics*, **32**,
    121–133.

Bibliography

Ahmad, S., Abdollahian, M., Zeephongsekul, P., and Abbasi, B.
    (2009), Multivariate non-normal process capability analysis,
    International Journal of Advanced Manufacturing Technology,
    44, 757+.
Gentle, J.E. (2003), *Random Number Generation and Monte Carlo
    Methods*, 2nd edn., New York: Springer-Verlag.
Jackson, J.E. (1959), Quality control for several related variables,
    Technometrics, **1**, 4+.

Johnson, N.L. (1949), Bivariate distributions based on simple translation systems, *Biometrika*, **36**, 297–304.

Johnson, N.L., Kotz, S.L., and Balakrishnan, N. (2000), *Continuous Multivariate Distributions, Models and Applications*, Vol. 1, 2nd edn., New York: John Wiley & Sons.

Johnson, R.A. and Wichern, D.W. (2002), *Applied Multivariate Statistical Analysis*, Upper Saddle River, NJ: Prentice Hall.

Mason, R.L. and Young, J.C. (2002), *Multivariate Statistical Process Control with Industrial Applications*, Philadelphia, PA: SIAM.

Taam, W., Subbaiah, P., and Liddy, J.W. (1993), A note on multivariate capability indices, *Journal of Applied Statistics*, **20**, 339–351.

Wang, F.K. (2006), Quality evaluation of a manufactured product with multiple characteristics, *Quality and Reliability Engineering International*, **22**, 225–236.

Chapter 8

Sample Size Determination

The ability to calculate precise estimates of process capability depends on collecting a sufficiently large, representative sample from the population. All of the measures of quality, from the percentage of nonconforming items through indices such as C_{pk}, become more precise as the sample size n increases. This chapter considers the important question of determining how large n should be to be sure that an acceptably precise estimate of process capability will be obtained.

Suppose that an analyst wishes to estimate a quality parameter such as θ, the proportion of nonconforming items or events in a process. It is decided to collect a random sample of n items in such a way as to be representative of that process. Through some mechanism, an estimate $\hat{\theta}$ will be obtained. In determining how large n should be, two general approaches may be taken:

1. The analyst may consider the variability of the estimate and in particular the size of the confidence interval or confidence bound associated with that estimate. n may then be selected such that the error bounds on the

estimate are acceptably tight. For example, n might be selected such that a 95% confidence interval for θ is no wider than ±10% of its true value.

2. Two values of θ may be selected, one indicative of good quality and the other of bad quality. A hypothesis test may then be constructed, using one of the values as the null hypothesis and the second value as the alternative hypothesis. n would then be selected such that the probability of choosing correctly between the 2 hypotheses is acceptably large.

This chapter examines methods for determining adequate sample sizes for estimating several important quality measures.

8.1 Sample Size Determination for Attribute Data

Chapters 2 and 3 discussed methods for estimating quality based on examining items or events and characterizing them as conforming or nonconforming. In Chapter 2, n items were collected and each one classified as good or bad. Quality was then estimated by assuming that the number of nonconforming items followed a binomial distribution. In Chapter 3, data were collected over a sampling interval and the number of nonconforming events during that interval was tabulated. Quality was then estimated based on the assumption that the number of events followed a Poisson distribution.

8.1.1 Sample Size Determination for Proportion of Nonconforming Items

Chapter 2 considered the situation in which a random sample of n items was collected from a large population. If X equals the number of nonconforming items in the sample, then

X follows a binomial distribution characterized by a single parameter θ, the probability that an individual item will be nonconforming. The point estimate for θ is given by

$$\hat{\theta} = \frac{X}{n} \qquad (8.1)$$

and a $100(1 - \alpha)\%$ confidence interval for θ is given by

$$\left[\frac{v_1 F_{1-\alpha/2,v_1,v_2}}{v_2 + v_1 F_{1-\alpha/2,v_1,v_2}}, \frac{v_3 F_{\alpha/2,v_3,v_4}}{v_4 + v_3 F_{\alpha/2,v_3,v_4}} \right] \qquad (8.2)$$

where

$$v_1 = 2X, \quad v_2 = 2(n - X + 1), \quad v_3 = 2(X + 1), \quad v_4 = 2(n - X) \qquad (8.3)$$

$F_{p,v,w}$ represents the value of Snedecor's F distribution with v and w degrees of freedom that is exceeded with probability p. Note that the degrees of freedom of the F distribution depend on the sample size n. As n increases, the interval becomes tighter.

8.1.1.1 Specification of Error Bounds

One method for determining an acceptable sample size begins by postulating a likely value for θ, such as $\theta = 0.001$. If a sample of size n is collected, then the expected value of X is

$$E(X) = n\theta \qquad (8.4)$$

To estimate θ to within $\pm 20\%$, it is necessary to find the smallest n such that the interval in Equation 8.2 is no wider than $(0.0008, 0.0012)$. Although no simple formula exists to solve this problem, it is a simple matter for statistical software to find the value of n by a direct search.

Example 8.1 Sample Size Determination for Proportion Nonconforming

Suppose an analyst wishes to estimate the proportion of nonconforming items in a population by taking a sample of n items and counting the number of defectives. It is expected that the percentage of defective items will be around 0.1%, or 1 in 1000. Further, the analyst is willing to accept an estimate with 95% error bounds of ±20%. Table 8.1 shows that the smallest sample capable of achieving those bounds is $n = 115,125$. Unfortunately, very large samples are required to obtain precise estimates of small proportions.

8.1.1.2 Specification of Alpha and Beta Risks

A second commonly used method of determining sample size for estimating a binomial proportion is to specify null and alternative hypotheses, such as

$$H_0: \theta = 0.001$$
$$H_A: \theta = 0.002$$

The alpha and beta risks are then specified, where

$$\alpha = \text{Prob(reject } H_0 \text{ when } H_0 \text{ is true)}$$
$$\beta = \text{Prob(do not reject } H_0 \text{ when } H_A \text{ is true)}$$

Table 8.1 Sample Size Required to Estimate a Binomial Proportion Near $\theta = 0.001$ to within 20% with 95% Confidence

Sample Size Determination
Parameter to be estimated: Binomial parameter
Desired tolerance: ±20.0% when proportion = 0.001
Confidence level: 95.0%
The required sample size is $n = 115,125$ observations.

For any given sample size, α and β can be calculated using the binomial distribution. Statistical software can easily calculate the minimum value of n for which both risks are less than any specified values.

Example 8.1 (Continued)

Suppose an analyst wishes to test the null hypothesis that the number of nonconforming items in a population is no greater than 0.001. A sufficiently large sample is desired such that the probability of rejecting the hypothesis that $\theta = 0.001$ is no greater than 5% when it is true and such that the probability of rejecting that hypothesis when $\theta = 0.002$ is at least 90%. Table 8.2 shows that $n = 13{,}017$ items are required to achieve such performance. Note that a one-sided test has been used, since the analyst is not concerned about not rejecting H_0 when the proportion of nonconforming items is less than that specified by the null hypothesis.

The probability of rejecting a null hypothesis when it is false equals $1 - \beta$ and is called the *power* of the test. The power of a hypothesis test for a binomial proportion may be expressed

Table 8.2   Sample Size Required to Test H_0: $\theta = 0.001$ versus H_A: $\theta = 0.002$ with α = 5% and β = 10%

Sample Size Determination
Parameter to be estimated: Binomial parameter
Desired power: 90.0% for proportion = 0.001 versus proportion = 0.002
Type of alternative: Greater than
Alpha risk: 5.0%
The required sample size is $n = 13{,}017$ observations.

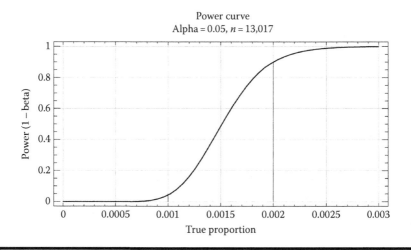

Figure 8.1   Power curve for test of H_0: $\theta = 0.001$ versus H_A: $\theta = 0.002$ with $\alpha = 5\%$ and $\beta = 10\%$.

as a function of the true proportion of nonconforming items in the population.

Example 8.1 (Continued)

The power curve, shown in Figure 8.1, passes through the points (0.001, 0.05) and (0.002, 0.90). The chance of rejecting the hypothesis that $\theta = 0.001$ clearly depends on how far the true proportion is from the null hypothesis.

8.1.2 Sample Size Determination for Rate of Nonconformities

Chapter 3 considered the problem of estimating the rate of nonconformities in a process when a sample of size n was collected. In that case, n could take either of two forms:

1. n could represent the number of discrete units examined, where each unit could have multiple nonconformities. A typical example is a sheet of glass, where the nonconformities are bubbles in the glass.

2. n could represent the size of a sampling interval during which unwanted events occurred. A typical example is a span of time in which unwanted events such as fatal aircraft accidents occurred.

The main parameter of interest in such cases is

$$\lambda = \text{Rate of nonconformities per unit}$$

If a sample of size n contains X nonconformities, then the rate of nonconformities is estimated by

$$\hat{\lambda} = \frac{X}{n} \tag{8.5}$$

A $100(1 - \alpha)\%$ confidence interval for λ is given by

$$\left[\frac{\chi^2_{1-\alpha/2,2X}}{2n}, \frac{\chi^2_{\alpha/2,2(X+1)}}{2n} \right] \tag{8.6}$$

In this case, X follows a Poisson distribution. As in the previous section, a sample size may be determined that either makes the interval in (8.6) acceptably tight or achieves the desired power for a set of hypotheses.

Example 8.2 Sample Size Determination for Rate of Nonconformities

Suppose an analyst wants to estimate the rate of intrusions into a computer network by monitoring that system. It is thought that the rate of intrusions is approximately $\lambda = 3$ intrusions per day. Management wishes to know how long they need to monitor the network so that the 95% upper bound on the estimated rate is no more than 120% of its true value.

To solve this problem, statistical software can easily search for the smallest n such that the upper confidence bound for λ is less than or equal to $3 * 1.2 = 3.6$ when $\hat{\lambda} = 3$. Table 8.3 shows that $n = 28$ days would be required to insure that the one-sided estimation error in λ is not more than 20% when the rate equals 3 per day.

Table 8.3   Sample Size Required to Achieve Upper Tolerance Bound of 20% When Rate Equals 3

Sample Size Determination
Parameter to be estimated: Poisson rate
Desired tolerance: 20.0% when rate = 3.0
Confidence level: 95.0%
The required sample size is $n = 28$ observations.

8.2 Sample Size Determination for Capability Indices

When the data available to assess quality consists of measurements rather than counts, quality indices such as C_{pk} are used to summarize process capability. This section considers the problem of determining adequate sample sizes for obtaining precise estimates of those indices.

8.2.1 Sample Size Determination for C_p and P_p

A commonly used index for measuring whether a process is capable of satisfying a specification consisting of both a lower specification limit (*LSL*) and an upper specification limit (*USL*) is the capability index defined by

$$C_p = \frac{USL - LSL}{6\sigma} \tag{8.7}$$

If the index is greater than or equal to 1.0, it means that the standard deviation σ is small enough that a range of $\mu \pm 3\sigma$ can fit completely within the specification limits. If the estimate of σ comes from a moving range or variation within subgroups, the index is usually referred to as C_p. If the estimate of σ includes all of the variation over the sampling period, the index is usually referred to as P_p, with the initial "P" standing for "Performance" rather than "Capability".

In Section 4.5.2, the $100(1 - \alpha)\%$ confidence interval for C_p was given as

$$\hat{C}_P \sqrt{\frac{\chi^2_{1-\alpha/2,\nu}}{\nu}} \leq C_P \leq \hat{C}_P \sqrt{\frac{\chi^2_{\alpha/2,\nu}}{\nu}} \qquad (8.8)$$

where ν is the degrees of freedom associated with the estimate of sigma used to compute the capability index. Table 4.6 lists the number of degrees of freedom associated with various estimates of σ. In each case, the degrees of freedom increases as the total number of observations increases.

Likewise, a one-sided lower bound for C_p may be expressed as

$$\hat{C}_P \sqrt{\frac{\chi^2_{1-\alpha,\nu}}{\nu}} \leq C_P \qquad (8.9)$$

Modifying Equation 8.9 by dividing each side by $\hat{C}_P$ gives

$$\frac{C_P}{\hat{C}_P} \geq \sqrt{\frac{\chi^2_{1-\alpha,\nu}}{\nu}} \qquad (8.10)$$

which converts the confidence bound from an absolute error to a bound for the relative error.

The most common method for selecting a sample size when estimating C_p or P_p is to specify the desired precision of the estimate. This requires specifying the desired level of confidence, such as 95%. Given this information, Equation 8.8 or 8.9 may be used to find the smallest degrees of freedom ν that will give the desired relative error, which may then be substituted into the equations in Table 4.6 to give a desired sample size n.

Example 8.3 Sample Size Determination for C_p

Suppose an analyst wishes to estimate the short-term capability index C_p by collecting n individual observations. After collecting the data, a 95% lower confidence bound for C_p will be calculated. The analyst wants to collect enough data so that the true index is no smaller than 90% of the calculated index with 95% confidence.

Table 8.4 shows that the smallest sample size that gives the desired precision is $n = 139$. The degrees of freedom $v = n - 1$, meaning that the ratio of the true index to the estimated index will be

$$\frac{C_P}{\hat{C}_P} \geq \sqrt{\frac{\chi^2_{1-\alpha,138}}{138}} \tag{8.11}$$

Since the value of the chi-square distribution with 138 degrees of freedom that is exceeded with probability 0.95 equals 111.875,

$$\frac{C_P}{\hat{C}_P} \geq 0.9004 \tag{8.12}$$

Figure 8.2 displays the minimum sample sizes needed to be 90%, 95%, and 99% confident that the ratio of the true capability index C_p to the estimated capability index $\hat{C}_p$ equals

Table 8.4   Sample Size Required to Be 95% Confident That C_p Is at Least 90% of That Estimated

Sample Size Determination (Capability Indices)
Capability index: C_p
Relative error: 10.0%
Confidence level: 95.0%
The required sample size is 139.

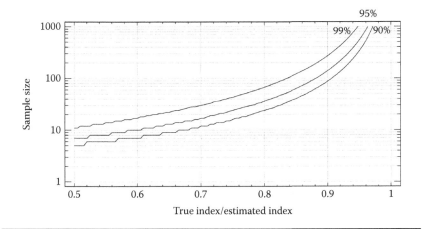

Figure 8.2   Required sample sizes for achieving various relative errors when constructing a lower confidence bound for C_p.

or exceeds various values. It may be seen that the required sample size increases quite dramatically for ratios in excess of 0.9. For example, to reduce the relative error of a 95% lower bound from 10% to 5% (corresponding to an increase of the ratio plotted on the horizontal axis from 0.9 to 0.95) increases the required sample size from 139 to approximately 550.

8.2.2 Sample Size Determination for C_{pk} and P_{pk}

A second common index that is used to estimate process capability, given either one- or two-sided specification limits, is C_{pk}. C_{pk} measures the distance from the process mean to the nearer specification limit divided by 3σ:

$$C_{pk} = \min\left[\frac{\mu - LSL}{3\sigma}, \frac{USL - \mu}{3\sigma}\right] \tag{8.13}$$

As with C_p, the index is labeled either C_{pk} or P_{pk} depending upon whether the estimate of σ measures short-term capability or long-term performance.

Section 4.5.4 gives equations for calculating 95% confidence intervals and bounds for C_{pk}. As in the previous section, the one-sided bound may be rewritten in terms of relative error as

$$\frac{C_{pk}}{\hat{C}_{pk}} \geq \left[1 - Z_\alpha \sqrt{\frac{1}{9n\hat{C}_{pk}^2} + \frac{1}{2v}} \right] \tag{8.14}$$

Since the right-hand side of Equation 8.14 becomes closer and closer to 1 as the sample size increases, the ratio may be made as close to 1 as desired by increasing n.

Example 8.4 Sample Size Determination for C_{pk}

Suppose an analyst wishes to estimate C_{pk} so that the relative error is no more than 10% when $\hat{C}_{pk} = 1.33$. Table 8.5 shows that this may be achieved by selecting a sample of 154 observations.

Figure 8.3 displays the minimum sample sizes needed to be 90%, 95%, and 99% confident that the ratio of the true capability index C_{pk} to the estimated capability index $\hat{C}_{pk}$ equals or exceeds various values.

Table 8.5 Sample Size Required to Be 95% Confident That C_{pk} Is at Least 90% of That Estimated

Sample Size Determination (Capability Indices)
Capability index: C_{pk}
Estimate: 1.33
Relative error: 10.0%
Confidence level: 95.0%
The required sample size is 154.

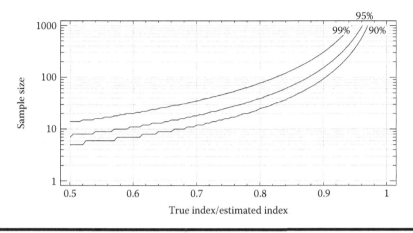

Figure 8.3 **Required sample sizes for achieving various relative errors when constructing a lower confidence bound for C_{pk}.**

8.3 Sample Size Determination for Statistical Tolerance Limits

Chapter 6 described the calculation of *statistical tolerance limits*. Given a set of measurements on n items randomly sampled from a population, statistical tolerance limits bound a specified proportion of the population from which the sample was taken at a given level of confidence. One of the best known methods for determining an adequate sample size for estimating statistical tolerance limits is due to Faulkenberry and Daly (1970). If the goal is to create tolerance limits for $P\%$ of the population, they suggest selecting a value of $P^* > P$ and choosing n so that the probability that the tolerance limits contain $P^*\%$ or more of the population is small. In the context of process capability analysis, their approach suffers from two drawbacks:

1. It assumes that the population from which the measurements are drawn is normal.
2. It is not connected in any way to the specification limits.

A criteria for selecting n that is more closely tied to the specification limits and does not require that the sample come from a normal distribution is as follows:

For two-sided specification limits: Given a probability distribution for X, select n such that the probability of obtaining a $100(1 - \alpha)\%$ tolerance interval for $P\%$ of the population that lies completely within the specification limits is greater than or equal to q.

For a one-sided upper specification limit: Given a probability distribution for X, select n such that the probability of obtaining a $100(1 - \alpha)\%$ upper tolerance bound for $P\%$ of the population that lies completely at or below the specification limit is greater than or equal to q.

For a one-sided lower specification limit: Given a probability distribution for X, select n such that the probability of obtaining a $100(1 - \alpha)\%$ lower tolerance bound for $P\%$ of the population that lies completely at or above the specification limit is greater than or equal to q.

To determine n, the analyst must specify

- The probability distribution for X, with values for each parameter
- The percentage of the population P for which the tolerance limits apply
- The level of confidence $100(1 - \alpha)\%$
- The desired probability q that the tolerance limits obtained from the data satisfy the specification limits

Solving this problem is best accomplished using a Monte Carlo simulation. The simulation proceeds as follows:

Step 1: Set $n = 4$.
Step 2: Generate m random samples of size n from the assumed distribution. For each random sample, calculate the statistical tolerance interval or bound. Tabulate the

proportion of tolerance intervals or bounds that satisfy the specification limits.

Step 3: If the proportion of randomly generated tolerance limits that satisfy the specs in Step 2 is greater than or equal to q, set the required sample size to n and stop.

Step 4: If the proportion of randomly generated tolerance limits that satisfy the specs is less than q, add 1 to n and return to Step 2.

It has been found that for practical problems, a value of m = 50,000 yields a sample size that does not vary much from one simulation to the next.

Example 8.5 Sample Size Determination for Statistical Tolerance Limits

Throughout this book, a sample of n = 100 medical device diameters has been used to illustrate methods for estimating process capability. In Chapter 5, it was shown that the data could be well-modeled by a largest extreme value distribution with mode = 1.97962 and scale parameter = 0.0138082. In Chapter 6, a 95% tolerance interval was obtained for 99% of the population from which the data were taken.

Now consider the following problem: Suppose that the medical device diameters come from a largest extreme value distribution with mode = 2.0 (the target value) and scale parameter = 0.015 (slightly larger than the estimate from the sample). If it is decided to take another sample from the population and estimate a 95% statistical tolerance interval for 99% of the medical devices, how large a sample is needed to have a 90% chance that the tolerance interval will be completely within the specification limits of [1.9,2.1]?

Figure 8.4 shows a typical dialog box used to control such a simulation. The fields specify the following:

Distribution: The assumed probability distribution for X.
Mode and Scale: Specified values for the distribution parameters.
Type of Limits: Whether a two-sided tolerance interval or a one-sided tolerance bound is desired.

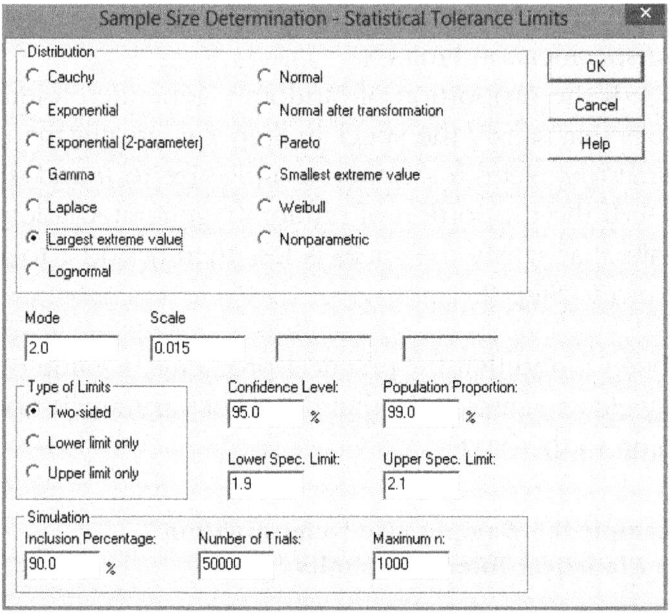

Figure 8.4 **Specification of parameters for determining sample size needed such that 90% of all 95–99 tolerance limits for the medical device data will be completely within spec.**

Confidence Level: Level of confidence for the tolerance limits.

Population Proportion: Percentage of the population to be bound by the tolerance limits.

Lower and Upper Spec Limits: The specification limits.

Inclusion Percentage: Desired percentage of time that the tolerance limits are to be completely within the specification limits.

Number of Trials: Number of random samples of size n used in each iteration of the simulation.

Maximum n: Maximum sample size considered. If the inclusion percentage is not met when n reaches this value, the simulation fails.

Table 8.6 shows that the required sample size is $n = 124$. For such a sample size, the tolerance interval for data with the exact parameters specified in Figure 8.4 would range from 1.97031 to 2.0915. This is tighter than the specification

Table 8.6 Required Sample Size for Estimating Statistical Tolerance Limits for Medical Device Data Given Specifications in Figure 8.4

Sample Size Determination (Statistical Tolerance Intervals)						
Conf. Level	*Pop. Percentage*	*Distri-bution*	*Mode*	*Scale*	*Lower Spec*	*Upper Spec*
95.0%	99.0%	Largest extreme value	2.0	0.015	1.9	2.1
Inclusion Percentage: 90%						
The required sample size is 124.						
Lower Tolerance Limit			*Upper Tolerance Limit*			
1.97031			2.0915			
Percentage of intervals within specification limits: 90.15%						

limits, which allows for random variability in the estimated mode and scale parameters. Note also that for samples of size $n = 124$, 90.15% of all simulated tolerance intervals were within the specification limits.

Note: It is not always possible to achieve the inclusion percentage specified, particularly if the distribution is very wide compared to the specification limits.

Reference

Faulkenberry, G.D. and Daly, J.C. (1970), Sample size for tolerance limits on a normal distribution, *Technometrics*, **12**, 813–821.

Bibliography

Gentle, J.E., Hardle, W., and Mori, Y. (2004), *Handbook of Computational Statistics: Concepts and Methods*, New York: Springer-Verlag.

Kramer, H.C. and Blasey, C.M. (2016), *How Many Subjects?: Statistical Power Analysis in Research*, 2nd edn., Newbury Park, CA: Sage Publications.

Law, A.M. (2015), *Simulation Modeling and Analysis*, 5th edn., New York: McGraw-Hill.

Chapter 9

Control Charts for Process Capability

The previous chapters of this book have concentrated on estimating process quality. The amount of data required to get a precise estimate of quantities such as the proportion of defective items or an index such as C_{pk} can be quite large. When the analyst needs to demonstrate that a process is capable of meeting specification limits or other requirements, there may be no way around taking a large sample.

Once a process has been shown to be capable, continued sampling of the process is necessary to demonstrate that the quality of the process has not changed significantly. For example, a medical device manufacturer might decide to sample each lot of devices produced. For economic reasons, the samples taken from each lot may need to be much smaller than the sample used to demonstrate initial compliance with the specifications. Taking smaller samples also makes logical sense, since the analyst begins with the belief that the process is capable and is most concerned with detecting any shifts from established quality levels.

This chapter considers two types of control charts that are useful for monitoring conformance with specification limits:

1. *Capability control charts*: These charts plot indices such as C_{pk} against established standard values with the goal of generating alerts whenever new data is inconsistent with those standards.

2. *Acceptance control charts*: These charts may be used to monitor processes with a very high C_{pk}, where it is not necessary to achieve a perfectly stable process. On such a chart, the control limits are placed far enough inside the specification limits to insure that a signal is generated whenever the short-term process mean comes too close to the specs, but otherwise the process is allowed to vary as it pleases around the long-term process mean.

9.1 Capability Control Charts

Capability control charts are used to monitor performance of a process after it has been deemed to be capable of meeting a set of requirements or specifications. The primary goal of such charts is to alert those in charge when a process appears to be moving away from its established levels. The charts are a type of Phase II statistical process control chart, commonly used to monitor processes in real time.

Figure 9.1 shows an example of a capability control chart. It consists of the following elements:

1. Point symbols showing the value of a capability index obtained from consecutive samples.
2. A centerline at the target value of the index, established from previous analyses.
3. Upper and lower control limits positioned above and below the centerline.

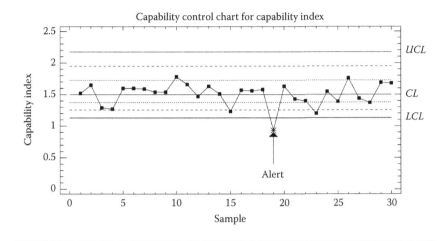

Figure 9.1 Capability control chart with control and warning limits.

4. Warning limits positioned 1/3rd and 2/3rds of the distance between the centerline and the control limits. The warning limits closest to the centerline are called the *inner warning limits*, while the warning limits farthest from the centerline are called the *outer warning limits*.

Alerts are generated whenever an individual index is either below the lower control limit or above the upper control limit. If desired, one or the other of the control limits may be omitted, usually the one corresponding to better than expected quality, since the main purpose of the chart is to detect situations in which quality is worse than expected.

The control limits are located at a distance from the centerline such that the probability of an estimated index falling beyond those limits when the process is operating at its expected level is small. The probability of such an alarm α is referred to as the *alpha risk* or *false alarm rate*. α is often set to the value associated with "3-sigma" control charts, for which $\alpha = 0.0027$ for a two-sided chart and $\alpha = 0.00135$ for a one-sided control chart.

Alerts may also be generated based on unusual patterns in the chart. The well-known Western Electric (WECO) runs rules may be applied, which generate alerts whenever

1. Any single point falls outside of the control limits
2. 2 out of 3 consecutive points fall beyond the outer warning limits, on the same side of the centerline
3. 4 out of 5 consecutive points fall beyond the inner warning limits, on the same side of the centerline
4. K consecutive points fall on the same side of the centerline, where K usually takes a value between 7 and 9

There are also a set of supplemental rules that are sometimes used to generate alerts:

1. K points in a row increasing or decreasing, where K usually takes a value between 6 and 8. Violation of this rule would indicate a possible trend.
2. 15 points in a row within the inner warning limits. Such a pattern would indicate better than expected performance.
3. 14 points in a row alternating direction (up and down). Such a pattern could indicate that the process is being constantly tweaked, which can add additional unwanted variability.
4. 8 points in a row beyond the inner warning limits, but not necessarily on the same side of the centerline.

Adding runs rules increases the chance that a change in the process will be detected quickly. However, it also increases the overall false alarm rate of the chart.

When implementing a capability control chart, it is important to examine the operating characteristic or OC curve associated with the chart. The OC curve plots the probability of not generating an alert when plotting any given sample as a function of the true value of the index. Figure 9.2 shows a typical OC curve for a two-sided control chart. It may be seen that the probability of not generating an alert is very large

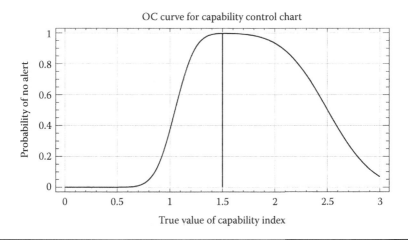

Figure 9.2 Operating characteristic curve for capability control chart.

at the target value but decreases as the true capability index moves above or below the target value.

A third curve of interest is called the ARL or *Average Run Length* curve. This curve plots the average number of samples before an alert is generated if the true capability index suddenly shifts from its assumed value to some other value. Small shifts take a long time to detect. Large shifts, on the other hand, will be detected much more quickly on average (Figure 9.3).

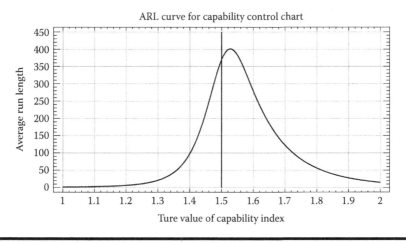

Figure 9.3 Average run length curve for capability control chart.

The rest of this section examines in detail capability control charts for various quality indices.

9.1.1 Control Chart for Proportion of Nonconforming Items

Chapter 2 considered the situation in which a random sample of n items was collected from a large population. If X equals the number of nonconforming items in the sample, then X follows a binomial distribution characterized by a single parameter θ, the probability that an individual item will be nonconforming. The point estimate for θ is given by

$$\hat{\theta} = \frac{X}{n} \qquad (9.1)$$

Given an assumed value for the proportion of nonconforming items θ_0, X follows the binomial distribution

$$p(x|\theta_0) = \binom{n}{x} \theta_0^x (1 - \theta_0)^{n-x} \qquad (9.2)$$

Example 9.1 Capability Control Chart for Proportion of Nonconforming Items

Table 9.1 displays a set of counts of nonconforming items, obtained from 30 samples of $n = 300$ items each. A lengthy study of the process has shown that it tends to produce approximately 1% defective items. A capability control chart is desired that has a false alarm probability of $\alpha = 0.0027$.

To create a capability control chart for θ, the centerline is placed at the expected value of $\hat{\theta}$:

$$CL = E\left[\frac{X}{n}\right] = \frac{1}{n} E[X] = \frac{1}{n} n\theta_0 = \theta_0 \qquad (9.3)$$

Table 9.1 Counts of Nonconforming Items from 30 Consecutive Samples of 300 Items Each

1	1	4	2	2	4
10	4	2	5	2	4
0	3	3	3	1	3
4	3	4	5	5	4
2	2	2	3	1	5

The upper control limit is located at the smallest value of X/n for which

$$\sum_{j=0}^{X} p(j|\theta_0) \geq 1 - \frac{\alpha}{2} \qquad (9.4)$$

where α is the false alarm probability of the chart, assuming a two-sided chart with no runs rules. For charts with only 1 control limit, the right-hand side of (9.4) is set equal to $1 - \alpha$. Likewise, the lower control limit is located at the smallest value of X/n for which

$$\sum_{j=0}^{X} p(j|\theta_0) > \frac{\alpha}{2} \qquad (9.5)$$

These equations assume that the consecutive estimates of $\hat{\theta}$ all come from samples of common size n. If the sizes of the samples vary, two choices are possible:

1. If the sample sizes do not vary much, then Equations 9.4 and 9.5 may be solved using the average sample size $\bar{n}$.
2. Control limits may be plotted for each sample using the separate sample sizes n_i, in which case the control limits look like step functions.

Example 9.1 (Continued)

To create a control chart for the sample data, the center line is set at $CL = 0.01$. To construct an upper control limit, Equation 9.4 is used to evaluate cumulative probabilities for the binomial distribution. The cumulative probabilities for a binomial distribution with $\theta = 0.01$ and $n = 300$ are given in Table 9.2. If the upper control limit is set at $X = 9$, then the false alarm probability of exceeding the UCL will equal $1 - 0.998977 = 0.001023$, which is less than $\alpha/2$. The upper control limit for the proportion of nonconforming items will therefore be set at

$$UCL = \frac{9}{300} = 0.03. \qquad (9.6)$$

To construct the lower control limit using Equation 9.5, note from Table 9.2 that the cumulative probability

Table 9.2 Binomial Distribution with $\theta = 0.01$ and $n = 300$

X	p(X)	F(X)
0	0.049041	0.049041
1	0.148609	0.197650
2	0.224414	0.422904
3	0.225170	0.647234
4	0.168877	0.816111
5	0.100985	0.917097
6	0.050153	0.967249
7	0.021277	0.988526
8	0.007871	0.996398
9	0.002580	0.998977
10	0.000758	0.999735

$F(X = 0 | \theta = 0.01, n = 300) = 0.049$, which is greater than $\alpha/2$. Thus the lower control limit must be set at LCL = 0, effectively making the chart one-sided on the upper side. Since the purpose of the chart is to detect situations that are worse than expected, a one-sided chart is acceptable.

Example 9.1 (Continued)

Figure 9.4 shows a capability control chart for the sample data. The seventh plotted value exceeds the upper limit of the chart, indicating an unusually large number of nonconformities in that sample. If created in real-time, the chart would have alerted process managers that an unusual event had just occurred at that point, quite possibly causing them to take action or perhaps to take a larger sample to determine whether or not a real problem existed.

Figure 9.5 shows the OC curve associated with the capability control chart for this data. The OC curve plots the probability of not generating an alert as a function of the true proportion of nonconforming items. Since the chart is

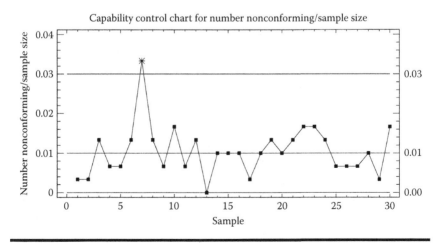

Figure 9.4   Capability control chart for proportion of nonconforming items.

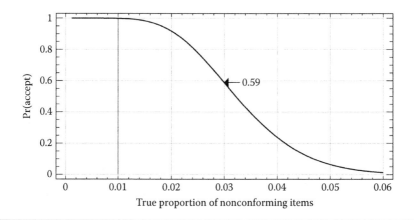

Figure 9.5 Operating characteristic curve for capability control chart for proportion of nonconforming items.

one-sided, the probability of no alert decreases with increasing values of θ. Note that the probability of no alert equals approximately 59% near $\theta = 0.03$. Higher alert probabilities could be obtained if necessary by increasing the sample size above $n = 300$. Sample size selection for control charts is discussed later in this chapter.

Example 9.1 (Continued)

The ARL at any given value of θ is related to the probability of not getting an alert when a sample is taken. In particular,

$$ARL(\theta) = \frac{1}{Prob(alert|\theta)} \qquad (9.7)$$

Consequently, the ARL at $\theta = 0.03$ is approximately equal to $1/(1 - 0.59) = 2.44$. This implies that if the proportion of nonconforming items in the process shifts suddenly from 1% to 3%, it will take on average about 2 and one-half sampling periods until an alert is generated by the chart. Increasing the sample size would reduce this response time.

9.1.2 *Control Chart for Rate of Nonconformities*

Chapter 3 considered the problem of estimating the rate of nonconformities in a process. If a sample of size n is collected and X nonconformities are observed, then the rate of nonconformities is estimated by

$$\hat{\lambda} = \frac{X}{n} \tag{9.8}$$

Given an assumed value for the rate of nonconformities λ_0, X follows the Poisson distribution:

$$p\left(x|\lambda_0\right) = \frac{\left(\lambda_0 n\right)^x \exp\left(-\lambda_0 n\right)}{x!} \tag{9.9}$$

To create a capability control chart for λ, the centerline is placed at the expected value of $\hat{\lambda}$:

$$CL = E\left[\frac{X}{n}\right] = \frac{1}{n}E\left[X\right] = \frac{1}{n}n\lambda_0 = \lambda_0 \tag{9.10}$$

As with the proportion of defective items, the upper control limit is set to the smallest value of X/n for which the cumulative probability equals or exceeds $1 - \alpha/2$, while the lower control limit is set to the smallest value that exceeds $\alpha/2$.

Example 9.2 Control Chart for Rate of Nonconformities

Table 1.2 contains data on the total number of accidents involving U.S. air carriers for each year between 1990 and 2014. Assuming a target level of $\lambda = 3.5$ accidents per million departures and setting $\alpha = 0.0027$ as before, Figure 9.6 shows a two-sided capability control chart for the accident rates on a yearly basis. The control limits are drawn using the actual number of departures each year,

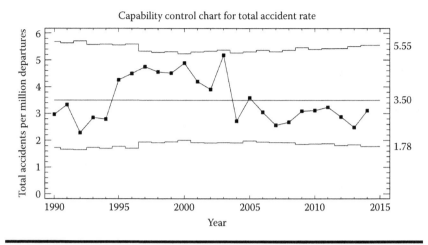

Figure 9.6 Capability control chart for U.S. air carrier accidents.

so they are not perfectly horizontal. Looking at the pattern of the points, it appears that the accident rate was relatively low until 1995. It is then relatively high between 1995 and 2003, after which it drops again. However, none of the values are far enough from 3.5 to be beyond the control limits.

Example 9.2 (Continued)

In order to generate alerts for small but persistent shifts away from target, it is useful to add the Western Electric rules. In Figure 9.7, alerts were generated for the following patterns:

1. 7 or more consecutive points on the same side of the centerline. This rule generated alerts in each year during the intervals 2001–2003 and 2012–2014.
2. Runs of 7 or more increasing or decreasing. This rule did not generate any alerts.
3. 4 out of 5 consecutive observations beyond the inner warning limits. This rule generated alerts each year during the interval 1998–2001 and again in 2003.
4. 2 out of 3 consecutive observations beyond the outer warning limits. This rule did not generate any alerts.

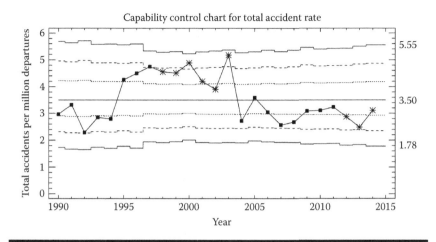

Figure 9.7   Capability control chart for U.S. air carrier accidents with runs rules violations marked.

The runs rules provide a strong signal that the accident rate has changed significantly over the years.

Addition of runs rules can dramatically reduce the ARL of the control charts when the change in the parameter is not large enough to generate individual points beyond the control limits.

9.1.3 Control Charts for C_p and P_p

The capability indices C_p and P_p are used to measure the capability of a process based on variable rather than attribute data. The indices compare the distance between a set of upper and lower specification limits to the variability in that process through the ratio

$$C_p = \frac{USL - LSL}{6\sigma} \tag{9.11}$$

Depending on the method used to estimate σ, the index may measure either short-term capability, in which case it is usually called C_p, or long-term performance, in which case it is called P_p.

Regardless of how σ is estimated, the distribution of its estimate follows that of a chi-square distribution, at least approximately. The difference among the estimates is in the degrees of freedom v associated with the chi-square reference distribution. Table 4.6 summarizes the degrees of freedom associated with various estimates.

To create a capability control chart for C_p or P_p, the analyst begins by specifying a target value $C_{p,0}$. This target value becomes the centerline of the chart. The upper control limit is then placed at

$$UCL = C_{p,0} \sqrt{\frac{v}{\chi^2_{1-\alpha/2,v}}} \qquad (9.12)$$

while the lower control limit is located at

$$LCL = C_{p,0} \sqrt{\frac{v}{\chi^2_{\alpha/2,v}}} \qquad (9.13)$$

Example 9.3 Control Chart for C_p

Suppose a process has been shown to be capable of performing at a level where $C_p = 2.0$. The process manager wishes to monitor the process to be sure that it continues to operate at that level. Consequently, at the end of each day a sample of $n = 30$ items is collected and used to calculate C_p. Table 9.3 shows a sequence of 25 such estimates.

Since each estimate of C_p is based on a sample of 30 measurements, the estimates $\hat{C}_p$ will follow a chi-square distribution with $v = 29$ degrees of freedom. Setting the false alarm rate $\alpha = 0.0027$, the upper control limit is thus given by

$$UCL = 2.0 \sqrt{\frac{29}{\chi^2_{.99865,29}}} = 2.0 \sqrt{\frac{29}{11.341}} = 3.198 \qquad (9.14)$$

Table 9.3   25 Consecutive Estimate of C_p, Each Based on $n = 30$ Observations

2.27	2.13	1.99	2.23	2.34
1.83	1.84	2.39	1.64	1.64
2.48	2.15	1.95	1.84	1.92
1.82	1.74	1.75	2.22	2.18
2.12	2.63	2.19	1.41	2.75
1.97	2.41	2.26	2.45	1.97

while the lower control limit is given by

$$LCL = 2.0\sqrt{\frac{29}{\chi^2_{.00135,29}}} = 2.0\sqrt{\frac{29}{57.225}} = 1.424 \qquad (9.15)$$

These are not very tight limits since there is a large amount of variability in a statistic such as C_p when it is calculated from only 30 measurements. However, Figure 9.8 shows that sample #24, which yielded a value $C_p = 1.41$, is below the lower control limit. An unusually low value of C_p

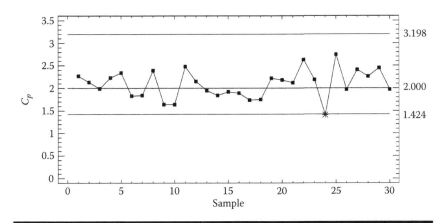

Figure 9.8   Capability control chart for 25 C_p estimates, each based on $n = 30$ observations.

indicates an unusually large estimate of σ. When such a signal occurs, it is good practice to take a larger sample from that period's production to determine whether the variability had actually increased or whether this value was a false alarm.

Note that the control limits for C_p are not symmetrically placed around the centerline, due to the skewed nature of the chi-square distribution.

The OC curve for a C_p or P_p control chart is calculated using the chi-square distribution.

Example 9.3 (Continued)

Figure 9.9 shows the OC curve for the capability control chart in Figure 9.8. Note that the probability of generating an alert is close to 50% when the true value of C_p is close to either of the control limits and drops quickly outside of the limits. An ARL chart can also be created as described earlier. Also, note that the OC curve is not symmetric due to the skewed nature of the chi-square distribution.

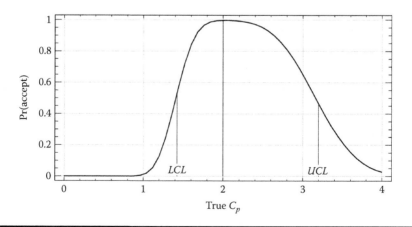

Figure 9.9   Operating characteristic curve for C_p control chart.

9.1.4 Control Charts for C_{pk} and P_{pk}

Capability control charts can also be generated for C_{pk} and P_{pk}, which measure the distance from the mean to the nearer specification limit in multiples of 3 times the process sigma. If both upper and lower specification limits are present, an index may be defined for each limit according to

$$C_{pk(lower)} = \frac{\mu - LSL}{3\sigma} \tag{9.16}$$

$$C_{pk(upper)} = \frac{USL - \mu}{3\sigma} \tag{9.17}$$

The combined index C_{pk} is then the smaller of the two indices (assuming both are calculated):

$$C_{pk} = \min\left[C_{pk(lower)}, C_{pk(upper)}\right] \tag{9.18}$$

To create a capability control chart for C_{pk} or P_{pk}, the analyst begins by specifying a target value $C_{pk,0}$. This target value becomes the centerline of the chart. To find the upper control limit, the following equation is solved for UCL:

$$UCL = C_{pk,0}\left[1 - Z_{\alpha/2}\sqrt{\frac{1}{9n \cdot UCL^2} + \frac{1}{2v}}\right]^{-1} \tag{9.19}$$

where n is the total number of observations and v is the degrees of freedom used to estimate sigma. To find the lower control limit, the following equation is solved for LCL:

$$LCL = C_{pk,0}\left[1 + Z_{\alpha/2}\sqrt{\frac{1}{9n \cdot LCL^2} + \frac{1}{2v}}\right]^{-1} \tag{9.20}$$

Both equations are easy to solve numerically.

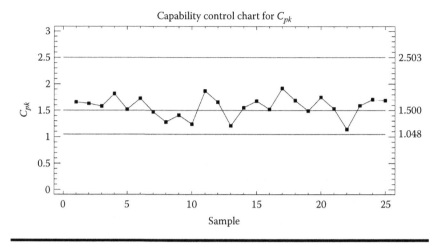

Figure 9.10   Capability control chart for 25 C_{pk} estimates, each based on n = 30 samples.

Example 9.4 Control Chart for C_{pk}

Suppose estimates of C_{pk} are generated from consecutive batches of a medical device manufacturing process, each based on a sample of n = 30 items. Let the target value $C_{pk,0}$ = 1.5 and let the false alarm probability α = 0.0027. Solving Equations 9.19 and 9.20 gives control limits of UCL = 2.503 and LCL = 1.048. Figure 9.10 shows the resulting capability control chart.

9.1.5 Sample Size Determination for Capability Control Charts

The OC curve may be used to select a sample size that is appropriate when creating a capability control chart for monitoring process capability. To select a sample size, the analyst must specify:

1. The parameter or index being monitored, such as the proportion of nonconforming items or the capability index C_{pk}
2. A target value for the parameter or index

3. Whether the control chart has both upper and lower limits or just one limit
4. A false alarm rate for the chart
5. An alternative value of the parameter or index
6. The desired power or ARL if the parameter being monitored suddenly shifts to the alternative value

Statistical software programs can easily search for the smallest number of observations n that yields a capability control chart with the desired characteristics.

Example 9.5 Sample Size Determination for C_{pk} Control Chart

Figure 9.11 shows the OC curve created while determining an adequate sample size for monitoring C_{pk}. The target value is set to $C_{pk} = 1.5$, while the alternative value is set to 1.0. The false alarm rate (probability of getting an alert when the target value is true) is set at $\alpha = 0.5\%$. The desired power (probability of getting an alert when the alternative is true) is set at $1 - \beta = 90\%$. The resulting OC curve

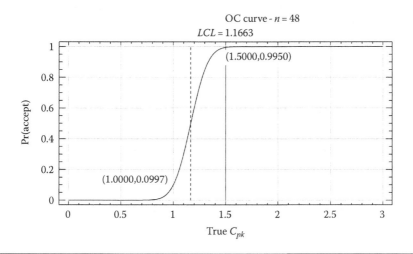

Figure 9.11 Operating characteristic curve for C_{pk} control chart with 90% power at $C_{pk} = 1.0$ and 0.5% false alarm rate at $C_{pk} = 1.5$, requiring samples of $n = 48$ observations.

shows that if $n = 48$ observations are collected each time a sample is taken, the probability of not getting an alert when the true $C_{pk} = 1.0$ is about 9.97%, which gives the desired power.

9.2 Acceptance Control Charts

For processes that operate at a high level of C_{pk}, strict control over the process may not be necessary. In particular, some variation in the process mean may be acceptable, provided the mean does not come too close to the specification limits. Montgomery (2013) describes how a standard X-bar chart, typically used to detect changes in a process mean, may be modified to control the proportion of nonconforming items. When used together with an R chart or S chart, the resulting *acceptance control chart* will detect changes in the process that correspond to unacceptable reductions in process capability.

Acceptance control charts are based on repeated sampling of the process. At any given time t, a sample of n items is collected. From each sample, statistics are calculated including the sample mean $\bar{x}_t$ and either the sample standard deviation s_t or the sample range R_t. Given specification limits USL and LSL and assuming a known process σ, upper and lower control limits are constructed that generate alerts whenever a sample mean gets too close to either of the specification limits. If only one specification limit is present, then a one-sided control chart is created instead.

There are two main approaches to constructing the control limits:

1. *Sigma multiple method*: This method specifies the target mean μ, the process standard deviation σ, the largest acceptable proportion of nonconforming items δ, and the desired false alarm probability α. The control

limits are established so that the probability of generating an alert is no larger than α whenever the probability of being beyond the specification limits is less than or equal to δ.

2. *Beta risk method*: This method specifies the target mean μ, the process standard deviation σ, an unacceptable proportion of nonconforming items γ, and the desired probability of getting an alert (1 − β) if the process mean shifts to a value for which the probability of being beyond the specification limits equals γ.

Note that the first method concentrates on controlling the false alarm probability, while the second method controls the power of the chart.

Example 9.6 Acceptance Control Charts

Suppose a process manager wants to monitor a process that produces medical devices. The diameter of the devices is required to fall within the range 2.0 ± 0.1 mm. Extensive studies of the process have established an operational standard with μ = 2.0 and σ = 0.01, which would put the specification limits 10 standard deviations from the mean. For such a process, the capability index C_p would be in excess of 3.

To monitor this process and detect situations when the process is no longer in control at the established mean and standard deviation, samples could be collected from the process periodically and the average diameters could be plotted on an *X*-bar control chart with control limits based on the established parameters. Figure 9.12 shows such a chart with 25 averages, each calculated from $n = 30$ values randomly generated from a normal distribution with μ = 2.0 and σ = 0.01. For this chart, the centerline is located at the assumed process mean. The upper control limit is located at

$$UCL = \mu + Z_{\alpha/2}\frac{\sigma}{\sqrt{n}} \qquad (9.21)$$

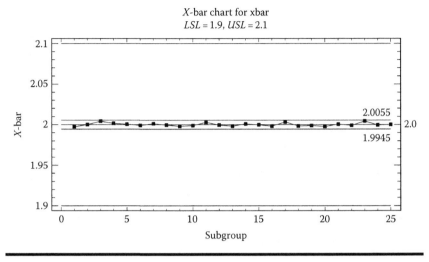

Figure 9.12   Phase II X-bar chart for medical device diameters.

and the lower control limit is at

$$LCL = \mu - Z_{\alpha/2} \frac{\sigma}{\sqrt{n}} \tag{9.22}$$

where $Z_{\alpha/2}$ is the value of the standard normal distribution that is exceeded with probability $\alpha/2$. Following standard practice and setting $\alpha = 0.0027$, then $Z_{\alpha/2} = 3$ and the control limits are located at $\pm 3\sigma/\sqrt{n}$.

It can be seen from the chart in Figure 9.12 that, for high C_{pk} processes, the control limits are far inside of the specification limits. While a sample average $\bar{x}_t$ beyond the control limits may signal that the mean and standard deviation of the process are not those assumed, it does not necessarily indicate that the process threatens to produce many nonconforming items. An acceptance chart increases the distance between the control limits, allowing the process mean to fluctuate as long as it does not get too close to the specification limits.

9.2.1 Sigma Multiple Method

To construct an acceptance chart using the sigma multiple method, the analyst begins by specifying the largest fraction of nonconforming items δ that is considered to be acceptable. Control limits are then established at positions that only rarely generate alerts unless the process mean moves close enough to the specification limits such that the fraction of nonconforming items is greater than δ. For the process to generate no more than δ nonconforming items, the lowest allowable value for the mean is

$$\mu_L = LSL + Z_\delta \sigma \tag{9.23}$$

and the highest allowable value for the mean is

$$\mu_U = USL - Z_\delta \sigma \tag{9.24}$$

The control limits are positioned far enough *outside* the means given above to give a false alarm rate equal to α if the mean moves to either μ_L or μ_U. Consequently, the limits for the chart of $\bar{x}_t$ are

$$UCL = USL - \left(Z_\delta - \frac{Z_\alpha}{\sqrt{n}} \right) \sigma \tag{9.25}$$

and

$$LCL = LSL + \left(Z_\delta - \frac{Z_\alpha}{\sqrt{n}} \right) \sigma \tag{9.26}$$

Example 9.6 (Continued)

Continuing the medical device example, suppose the process manager decides that the maximum acceptable fraction of nonconforming items is δ = 0.0001. The value of the standard normal distribution exceeded with probability 0.0001 is Z_δ = 3.719. This implies that the mean can be as great as μ_U = 2.0628 or as small as μ_L = 1.9372 without exceeding the specified proportion of nonconforming items.

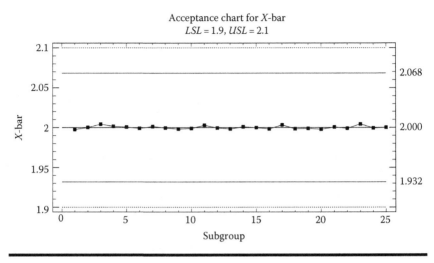

Figure 9.13 Acceptance control chart constructed using sigma multiple method.

The control limits are located at a distance $(3.719 - 3/\sqrt{30})\sigma$ inside of the specification limits, as shown in Figure 9.13.

Figure 9.14 shows the operating characteristic curve for the resulting chart. Notice that the probability of acceptance (not getting an alert) is large at all values between μ_L and μ_U, while it falls off rapidly beyond those values.

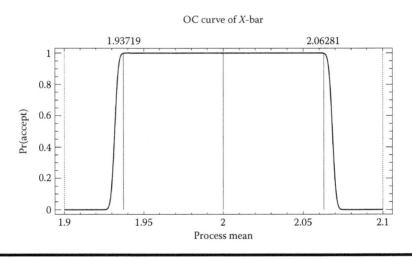

Figure 9.14 OC curve for acceptance control chart created using sigma multiple method.

9.2.2 Beta Risk Method

An alternative method for constructing an acceptance chart is to position the control limits such that the risk of *not* getting a signal equals β when the mean shifts to either μ_L or μ_U, where those upper and lower means correspond to a fraction of nonconforming items γ that is unacceptable. This places the control limits *inside* of μ_L and μ_U at

$$UCL = USL - \left(Z_\gamma + \frac{Z_\beta}{\sqrt{n}} \right) \sigma \qquad (9.27)$$

and

$$LCL = LSL + \left(Z_\gamma + \frac{Z_\beta}{\sqrt{n}} \right) \sigma \qquad (9.28)$$

Of course, controlling the missed alarm rate β impacts the false alarm rate α.

Example 9.6 (Continued)

Suppose a process manager wishes to construct a control chart that will generate a signal 95% of the time if the mean shifts to a location corresponding to 0.1% nonconforming items. Using the beta risk method with $\beta = 0.05$ and $\gamma = 0.001$, the control limits are located at a distance equal to $(3.09024 + 1.64486/\sqrt{30})\sigma$ inside of the specification limits, as shown in Figure 9.15. These limits are tighter than using the sigma multiple method, but still much wider than using a standard X-bar chart.

The difference in performance between the two methods of constructing an acceptance chart may be seen by comparing the OC curve in Figure 9.16 to that of Figure 9.14. The beta risk method insures that the probability of getting a signal is large at γ, while the sigma multiple method insures that the false alarm probability at δ is small.

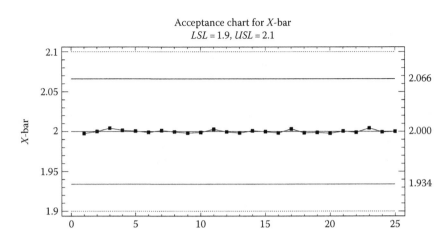

Figure 9.15 Acceptance control chart constructed using beta risk method.

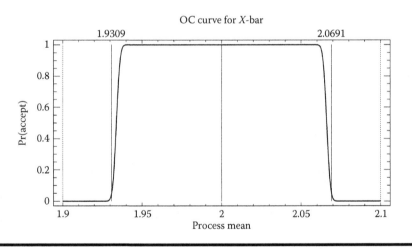

Figure 9.16 OC curve for acceptance control chart created using beta risk method.

Montgomery (2013) shows that the only way to control both the alpha and beta risks is to select the proper sample size n. In his example, two proportions are defined:

δ = fraction of nonconforming items that the manager is willing to accept with probability $1 - \alpha$

γ = fraction of nonconforming items that the manager wishes to reject with probability $1 - \beta$

An acceptance chart that meets both these conditions requires plotting subgroup means, each of which is based on

$$n = \left(\frac{Z_\alpha + Z_\beta}{Z_\delta - Z_\gamma} \right)^2 \qquad (9.29)$$

observations.

Example 9.7 Sample Size Determination for Acceptance Control Charts

Returning to Example 9.6, suppose the manager wished to adjust his sampling plan so that an acceptance control chart would have a 95% chance of generating an alert when the process was producing 0.1% or more nonconforming items, yet have only a 0.27% chance of generating an alert when the process was producing 0.01% defective items. Using Equation 9.29:

$$n = \left(\frac{Z_{.0027} + Z_{.05}}{Z_{.0001} - Z_{.001}} \right)^2 = \left(\frac{2.782 + 1.645}{3.719 - 3.090} \right)^2 = 49.5 \qquad (9.30)$$

Each time a sample is taken, 50 devices should be measured and the mean diameter plotted on the chart.

Reference

Montgomery, D.C. (2013), *Introduction to Statistical Quality Control*, 7th edn., Hoboken, NJ: John Wiley & Sons.

Bibliography

Carey, R.G. (2003), *Improving Healthcare with Control Charts: Basic and Advanced SPC Methods and Case Studies*, Milwaukee, WI: ASQ Quality Press.

Duncan, A.J. (1986), *Quality Control and Industrial Statistics*, 5th edn., Homewood, IL: Richard D. Irwin, Inc.

Relyea, D.B. (2011), *The Practical Application of the Process Capability Study: Evolving from Product Control to Process Control*, New York: Productivity Press.

Wheeler, D.J. (2015), *Advanced Topics in Statistical Process Control*, 2nd edn., SPC Press, Inc, Knoxville, TN.

Conclusion

This book has examined methods for estimating how capable a process is of meeting the specifications established for it. In all cases, primary interest has centered on estimating θ, the proportion of time that a product or service does not satisfy the specifications. Estimating capability has been accomplished by taking a sample of size n from the population and then using one of three methods:

1. Directly counting the number of nonconformities
2. Fitting a normal or nonnormal distribution to measurements made on each item and estimating the proportion of that distribution that is beyond the specification limits
3. Calculating statistical tolerance limits for the measurements

Considerable emphasis has been given to obtaining confidence limits for θ and other estimates of quality. In some cases, these limits may be obtained analytically. In other cases, bootstrapping and Monte Carlo methods must be used. As in all statistical analyses, providing measures of precision for the numbers that are calculated is critical for their proper interpretation.

Throughout the book, the problem of determining an adequate sample size has also been considered. One of the most frequently asked questions of statisticians is "How much data do I need?" The statistician's usual response is "How well do you want to know the answer?" This book has framed the sample size problem in terms that should be easy to use in practice. By specifying either the precision of the estimated quality index or the power of a test against a specific alternative, the required sample size may be calculated.

Much attention has also been given to being sure that the data used to estimate quality satisfy the assumptions of the methods applied. While much data follows a normal distribution, much does not. Assuming normality when it does not exist may yield attractive numerical results, but they mean nothing in reality. Selecting alternative distributions and finding normalizing transformations are important tools for making proper use of the methods described in this book.

Since quality rarely depends on a single characteristic, consideration has also been given to the analysis of multivariate capability. Considering multiple variables simultaneously is important, particularly when those variables are correlated. Unfortunately, current practice usually concentrates on only one variable at a time.

Finally, the methods presented in this book have illustrated the results of the capability analyses using graphical output wherever possible. Visualizing how well or poorly a process is performing helps bring capability statistics to life.

Appendix A: Probability Distributions

This appendix lists distributions that are useful for modeling discrete and continuous variables. Distributions for discrete variables are listed first.

Bernoulli Distribution

Range of X: 0 or 1

Common use: Representation of an event with two possible outcomes. In the distributions below, the primary outcome will be referred to as a "success".

PMF: $p(x)=p^x(1-p)^{1-x}$

Parameters: Event probability $0 \leq p \leq 1$

Mean: p

Variance: $p(1-p)$

Binomial Distribution

Range of X: 0, 1, 2, ..., n

Common use: Distribution of number of successes in a sample of n independent Bernoulli trials. Commonly used for number of defects in a sample of size n.

PMF: $p(x) = \binom{n}{x} p^x (1-p)^{n-x}$

Parameters: Event probability $0 \leq p \leq 1$, number of trials $n \geq 1$

Mean: np

Variance: $np(1 - p)$

Discrete Uniform Distribution

Range of X: $a, a + 1, a + 2, \ldots, b$

Common use: Distribution of an integer-valued variable with both lower bound and upper bound.

PMF: $p(x) = \dfrac{1}{b - a + 1}$

Parameters: Lower limit a, upper limit $b \geq a$

Mean: $\dfrac{a + b}{2}$

Variance: $\dfrac{(b - a + 1)^2 - 1}{12}$

Geometric Distribution

Range of X: $0, 1, 2, \ldots$

Common use: Waiting time until the occurrence of the first success in a sequence of independent Bernoulli trials. Number of items inspected before the first defect is found.

PMF: $p(x) = p(1 - p)^x$

Parameters: Event probability $0 \leq p \leq 1$

Mean: $\dfrac{1 - p}{p}$

Variance: $\dfrac{1 - p}{p^2}$

Hypergeometric Distribution

Range of X: $\max(0, n - m), 1, 2, \ldots, \min(m, n)$

Common use: Number of items of a given type selected from a finite population with two types of items, such as good and bad. Acceptance sampling from lots of fixed size.

PMF: $p(x) = \dfrac{\dbinom{m}{x}\dbinom{N-m}{n-x}}{\dbinom{N}{n}}$

Parameters: Population size N, number of items $0 \le m \le N$, sample size n

Mean: $\dfrac{mn}{N}$

Variance: $\dfrac{\left(\dfrac{mn}{N}\right)\left(1-\dfrac{m}{N}\right)(N-n)}{(N-1)}$

Negative Binomial (Pascal) Distribution

Range of X: 0, 1, 2, …

Common use: Waiting time until the occurrence of k successes in a sequence of independent Bernoulli trials. Number of good items inspected before the kth defect is found.

PMF: $p(x) = \dbinom{x+k-1}{x} p^k (1-p)^x$

Parameters: Event probability p, number of successes k

Mean: $k\dfrac{(1-p)}{p}$

Variance: $\dfrac{k(1-p)}{p^2}$

Poisson Distribution

Range of X: 0, 1, 2, …

Common use: Number of events in an interval of fixed size when events occur independently. Common model for number of defects per unit.

PMF: $p(x) = \dfrac{\lambda^x e^{-\lambda}}{x!}$

Parameters: Mean $\lambda > 0$

Mean: λ

Variance: λ

Beta Distribution

Range of X: $0 \leq X \leq 1$

Common use: Distribution of a random proportion.

PDF: $f(x) = \dfrac{x^{\alpha_1 - 1}(1 - x)^{\alpha_2 - 1}}{B(\alpha_1, \alpha_2)}$

Parameters: Shape $\alpha_1 > 0$, shape $\alpha_2 > 0$

Mean: $\dfrac{\alpha_1}{\alpha_1 + \alpha_2}$

Variance: $\dfrac{\alpha_1 \alpha_2}{(\alpha_1 + \alpha_2)^2 (\alpha_1 + \alpha_2 + 1)}$

Beta Distribution (4-Parameter)

Range of X: $a \leq X \leq b$

Common use: Model for variable with both lower and upper limits. Often used as a prior distribution for Bayesian analysis.

PDF: $f(x) = \dfrac{(x - a)^{\alpha_1 - 1}(b - x)^{\alpha_2 - 1}}{B(\alpha_1, \alpha_2)(b - a)^{\alpha_1 + \alpha_2 - 1}}$

Parameter: Shape $\alpha_1 > 0$, shape $\alpha_2 > 0$, lower limit a, upper limit $b > a$

Mean: $a + \dfrac{b\alpha_1}{\alpha_1 + \alpha_2}$

Variance: $\dfrac{\alpha_1 \alpha_2 (b - a)^2}{(\alpha_1 + \alpha_2)^2 (\alpha_1 + \alpha_2 + 1)}$

Birnbaum-Saunders Distribution

Range of X: $X > 0$

Common use: Model for the number of cycles needed to cause a crack to grow to a size that would cause a fracture to occur.

PDF: $f(x) = \dfrac{\sqrt{\dfrac{x}{\theta}} + \sqrt{\dfrac{\theta}{x}}}{2\beta x} \phi\left(\dfrac{1}{\beta}\left(\sqrt{\dfrac{x}{\theta}} - \sqrt{\dfrac{\theta}{x}}\right)\right)$ where $\phi(z)$ is the standard normal pdf

Parameters: Shape $\beta > 0$, scale $\theta > 0$

Mean: $\theta\left(1 + \dfrac{\beta^2}{2}\right)$

Variance: $(\theta\beta)^2\left(1 + \dfrac{5\beta^2}{4}\right)$

Cauchy Distribution

Range of X: All real X

Common use: Model for measurement data with longer and flatter tails than the normal distribution.

PDF: $f(x) = \dfrac{1}{\pi\beta}\left[\left(\dfrac{x - \theta}{\beta}\right)^2 + 1\right]^{-1}$

Parameters: Mode θ, scale $\beta > 0$

Mean: Not defined

Variance: Not defined

Chi-Square Distribution

Range of X: $X \geq 0$

Common use: Distribution of the sample variance s^2 from a normal population.

PDF: $f(x) = \dfrac{x^{(v-2)/2}e^{-x/2}}{2^{v/2}\Gamma\left(\dfrac{v}{2}\right)}$

Parameters: Degrees of freedom $v > 0$

Mean: v

Variance: $2v$

Erlang Distribution

Range of X: $X \geq 0$

Common use: Length of time before α arrivals in a Poisson process.

PDF: $f(x) = \dfrac{\lambda^{\alpha} x^{\alpha-1} e^{-\lambda x}}{\Gamma(\alpha)}$

Parameters: Integer shape $\alpha \geq 1$, scale $\lambda > 0$

Mean: $\dfrac{\alpha}{\lambda}$

Variance: $\dfrac{\alpha}{\lambda^2}$

Exponential Distribution

Range of X: $X \geq 0$

Common use: Time between consecutive arrivals in a Poisson process. Lifetime of items with a constant hazard rate.

PDF: $f(x) = \lambda e^{-\lambda x}$

Parameters: Rate $\lambda > 0$

Mean: $\dfrac{1}{\lambda}$

Variance: $\dfrac{1}{\lambda^2}$

Exponential Distribution (2-Parameter)

Range of X: $X \geq \theta$

Common use: Model for lifetimes with a lower limit.

PDF: $f(x) = \lambda e^{-\lambda(x-\theta)}$

Parameters: Threshold θ, scale $\lambda > 0$

Mean: $\theta + \dfrac{1}{\lambda}$

Variance: $\dfrac{1}{\lambda^2}$

Exponential Power Distribution

Range of X: All real X

Common use: Symmetric distribution with parameter controlling the kurtosis. Special cases include normal and Laplace distributions.

PDF: $f(x) = \dfrac{1}{\Gamma\left(1 + \dfrac{1+\beta}{2}\right) 2^{1+(1+\beta)/2} \phi} \exp\left(-\dfrac{1}{2}\left|\dfrac{x-\mu}{\phi}\right|^{2/(1+\beta)}\right)$

Parameters: Mean μ, shape $\beta \geq -1$, scale $\phi > 0$

Mean: μ

Variance: $2^{(1+\beta)} \left\{ \dfrac{\Gamma\left[\dfrac{3}{2}(1+\beta)\right]}{\Gamma\left[\dfrac{1}{2}(1+\beta)\right]} \right\} \phi^2$

F Distribution

Range of X: $X \geq 0$

Common use: Distribution of the ratio of two independent variance estimates from a normal population.

PDF: $f(x) = \dfrac{\Gamma\left(\dfrac{v+w}{2}\right) v^{v/2} w^{w/2} x^{(v-2)/2}}{\Gamma\left(\dfrac{v}{2}\right)\Gamma\left(\dfrac{w}{2}\right)(w+vx)^{(v+w)/2}}$

Parameters: Numerator degrees of freedom $v > 0$, denominator degrees of freedom $w > 0$

Mean: $\dfrac{w}{w-2}$ if $w > 2$

Variance: $\dfrac{2w^2(v+w-2)}{v(w-2)^2(w-4)}$ if $w > 4$

Folded Normal Distribution

Range of X: $X \geq 0$

Common use: Absolute values of data that follow a normal distribution.

PDF: $f(x) = \dfrac{1}{\sigma}\sqrt{\dfrac{2}{\pi}}\cosh\left(\dfrac{\mu x}{\sigma^2}\right)e^{-\frac{x^2+\mu^2}{2\sigma^2}}$

Parameters: Location $\mu > 0$, scale $\sigma \geq 0$

Mean: $\sigma\sqrt{\dfrac{2}{\pi}}\exp\left(-\dfrac{\mu^2}{2\sigma^2}\right) - \mu\left[1 - 2\Phi\left(\dfrac{\mu}{\sigma}\right)\right]$ where $\Phi(z)$ is the standard normal cdf

Variance: $\mu^2 + \sigma^2 - \left[\sigma\sqrt{\dfrac{2}{\pi}}\exp\left(-\dfrac{\mu^2}{2\sigma^2}\right) + erf\left(\dfrac{\mu}{\sqrt{2}\sigma}\right)\mu\right]^2$

Gamma Distribution

Range of X: $X \geq 0$

Common use: Model for positively skewed measurements. Time to complete a task, such as a repair.

PDF: $f(x) = \dfrac{\lambda^\alpha(x - \theta)^{\alpha-1}e^{-\lambda x}}{\Gamma(\alpha)}$

Parameters: Shape $\alpha > 0$, scale $\lambda > 0$

Mean: $\dfrac{\alpha}{\lambda}$

Variance: $\dfrac{\alpha}{\lambda^2}$

Gamma Distribution (3-Parameter)

Range of X: $X \geq \theta$

Common use: Model for positively skewed data with a fixed lower bound.

PDF: $f(x) = \dfrac{\lambda^\alpha(x - \theta)^{\alpha-1}e^{-\lambda(x-\theta)}}{\Gamma(\alpha)}$

Parameters: Shape $\alpha > 0$, scale $\lambda > 0$, threshold θ

Mean: $\theta + \dfrac{\alpha}{\lambda}$

Variance: $\dfrac{\alpha}{\lambda^2}$

Generalized Gamma Distribution

Range of X: $X > 0$

Common use: General distribution containing the exponential, gamma, Weibull, and lognormal distributions as special cases.

PDF: $f(x) = \dfrac{\lambda}{\sigma x} \dfrac{1}{\Gamma(\lambda^{-2})}$

$$\times \exp\left[\frac{w}{\lambda} + \frac{\ln(\lambda^{-2})}{\lambda^2} - \exp(\lambda w + \ln(\lambda^{-2}))\right]$$

where $w = [\log(x) - \mu]/\sigma$

Parameters: Location μ, scale $\sigma > 0$, shape $\lambda > 0$

Mean: $\exp\left[\mu + \dfrac{\sigma}{\lambda}\ln(\lambda^{-2})\right] \dfrac{\Gamma\left(\dfrac{\sigma}{\lambda} + \lambda^{-2}\right)}{\Gamma(\lambda^{-2})}$

Variance: $\exp\left[\mu + \dfrac{\sigma}{\lambda}\ln(\lambda^{-2})\right]^2 \left[\dfrac{\Gamma\left(\dfrac{2\sigma}{\lambda} + \lambda^{-2}\right)}{\Gamma(\lambda^{-2})} - \dfrac{\Gamma^2\left(\dfrac{2\sigma}{\lambda} + \lambda^{-2}\right)}{\Gamma^2(\lambda^{-2})}\right]$

Generalized Logistic Distribution

Range of X: All real x

Common use: Used for the analysis of extreme values. May be either left-skewed or right-skewed, depending on the shape parameter.

PDF: $f(x) = \dfrac{\gamma}{\kappa} \dfrac{\exp(-(x-\mu)/\kappa)}{\left[1 + \exp(-(x-\mu)/\kappa)\right]^{1+\gamma}}$

Parameters: Location μ, scale $\kappa > 0$, shape $\gamma > 0$

Mean: $\mu + [0.57722 + \Psi(\gamma)]\kappa$ where $\Psi(z)$ is the digamma function

Variance: $\left[\dfrac{\pi^2}{6} + \Psi'(\gamma)\right]\kappa^2$

Half Normal Distribution

Range of X: $X \geq \mu$

Common use: Normal distribution folded about its mean.

PDF: $f(x) = \dfrac{1}{\sigma} \sqrt{\dfrac{2}{\pi}} \exp\left(-\dfrac{1}{2}\left(\dfrac{x-\mu}{\sigma}\right)^2\right)$

Parameters: Scale $\sigma > 0$, threshold μ

Mean: $\mu + \sqrt{\dfrac{2}{\pi}}\sigma$

Variance: $\sigma^2\left(1 - \dfrac{2}{\pi}\right)$

Inverse Gaussian Distribution

Range of X: $X > 0$

Common use: First passage time in Brownian motion.

PDF: $f(x) = \sqrt{\dfrac{\beta}{2\pi\theta x^3}} \exp\left(\dfrac{-\beta(x-\theta)^2}{2x\theta^3}\right)$

Parameters: Mean $\theta > 0$, scale $\beta > 0$

Mean: θ

Variance: $\dfrac{\theta^2}{\beta}$

Johnson SB Distribution

Range of X: $\theta < X < \theta + \lambda$

Common use: Bounded part of Johnson family, used for modeling nonnormal data.

PDF: $f(x) = \dfrac{\partial}{\lambda\sqrt{2\pi}} \dfrac{1}{y(1-y)} e^{-\frac{1}{2}\left(\gamma+\delta\ln\left(\frac{y}{1-y}\right)\right)^2}$ where $y = (x-\theta)/\lambda$

Parameters: Location θ, scale $\lambda > 0$, shape γ shape $\delta > 0$

Johnson SL Distribution

Range of X: $X > \theta$

Common use: Unbounded part of Johnson family, used for modeling nonnormal data.

PDF: $f(x) = \dfrac{\partial}{\lambda\sqrt{2\pi}} \dfrac{1}{y} e^{-\frac{1}{2}(\gamma+\delta\ln(y))^2}$ where $y = (x-\theta)/\lambda$

Parameters: Location θ, shape γ shape δ > 0

Mean: $\theta + \exp\left(\dfrac{\delta^{-2}}{2}\right)\exp\left(-\dfrac{\gamma}{\delta}\right)$

Variance: $\exp(2\delta^{-2})\exp(-2\gamma/\delta)$

Johnson SU Distribution

Range of X: All real X

Common use: Bounded part of Johnson family, used for modeling nonnormal data.

PDF: $f(x) = \dfrac{\partial}{\lambda\sqrt{2\pi}}\dfrac{1}{\sqrt{1+y^2}}\, e^{-\frac{1}{2}\left(\gamma + \delta\sinh^{-1}(y)\right)^2}$ where

$y = (x - \theta)/\lambda$

Parameters: Location θ, scale λ > 0, shape γ shape δ > 0

Mean: $\theta - \lambda\exp\left(\dfrac{\delta^{-2}}{2}\right)\sinh\left(\dfrac{\gamma}{\delta}\right)$

Variance: $\dfrac{\lambda^2}{2}\left(\exp\left(\delta^{-2}\right) - 1\right)\left(\exp\left(\delta^{-2}\right)\cosh\left(\dfrac{2\gamma}{\delta}\right) + 1\right)$

Laplace (Double Exponential) Distribution

Range of X: All real X

Common use: Symmetric distribution with a very pronounced peak and long tails.

PDF: $f(x) = \dfrac{\lambda}{2}e^{-\lambda|x-\mu|}$

Parameters: Mean μ, scale λ > 0

Mean: μ

Variance: $\dfrac{2}{\lambda^2}$

Largest Extreme Value Distribution

Range of X: All real X

Common use: Distribution of the largest value in a sample from many distributions. Also used for positively skewed measurement data.

$$\text{PDF: } f(x) = \frac{1}{\beta} \exp\left\{-\left(\frac{x-\gamma}{\beta}\right) - \exp\left(-\frac{x-\gamma}{\beta}\right)\right\}$$

Parameters: Mode γ, scale $\beta > 0$

Mean: $\gamma + 0.57722\,\beta$

Variance: $\dfrac{\beta^2\pi^2}{6}$

Logistic Distribution

Range of X: All real X

Common use: Used as a model for growth and as an alternative to the normal distribution.

$$\text{PDF: } f(x) = \frac{1}{\sigma} \frac{\exp(z)}{\left[1+\exp(z)\right]^2} \quad \text{where } z = \frac{x-\mu}{\sigma}$$

Parameters: Mean μ, standard deviation $\sigma > 0$

Mean: μ

Variance: σ^2

Loglogistic Distribution

Range of X: $X > 0$

Common use: Used for data where the logarithms follow a logistic distribution.

$$\text{PDF: } f(x) = \frac{1}{\sigma x} \frac{\exp(z)}{\left[1+\exp(z)\right]^2} \quad \text{where } z = \frac{\ln(x)-\mu}{\sigma}$$

Parameters: Median $\exp(\mu)$, shape $\sigma > 0$

Mean: $\exp(\mu)\Gamma(1+\sigma)\Gamma(1-\sigma)$

Variance: $\exp(2\mu)[\Gamma(1+2\sigma)\Gamma(1-2\sigma) - \Gamma^2(1+\sigma)\Gamma^2(1-\sigma)]$

Loglogistic Distribution (3-Parameter)

Range of X: $X > \theta$

Common use: Used for data where the logarithms follow a logistic distribution after subtracting a threshold value.

$$\text{PDF: } f(x) = \frac{1}{\sigma x} \frac{\exp(z)}{\left[1+\exp(z)\right]^2} \quad \text{where } z = \frac{\ln(x-\theta)-\mu}{\sigma}$$

Parameters: Median exp(μ), shape $\sigma > 0$, threshold θ
Mean: $\theta + \exp(\mu)\Gamma(1+\sigma)\Gamma(1-\sigma)$
Variance: $\exp(2\mu)[\Gamma(1+2\sigma)\Gamma(1-2\sigma) - \Gamma^2(1+\sigma)\Gamma^2(1-\sigma)]$

Lognormal Distribution
Range of X: $X > 0$
Common use: Used for data where the logarithms follow a normal distribution.

PDF: $f(x) = \dfrac{1}{x\sqrt{2\pi}\sigma} e^{-\frac{(\ln x - \mu)^2}{2\sigma^2}}$

Parameters: Location μ, scale $\sigma > 0$
Mean: $e^{\mu + \sigma^2/2}$
Variance: $e^{2\mu + \sigma^2}\left(e^{\sigma^2} - 1\right)$

Lognormal Distribution (3-Parameter)
Range of X: $X > \theta$
Common use: Used for data where the logarithms follow a normal distribution after subtracting a threshold value.

PDF: $f(x) = \dfrac{1}{(x-\theta)\sqrt{2\pi}\sigma} e^{-\frac{(\ln(x-\theta)-\mu)^2}{2\sigma^2}}$

Parameters: Location μ, scale $\sigma > 0$, threshold θ
Mean: $\theta + e^{\mu + \sigma^2/2}$
Variance: $e^{2\mu + \sigma^2}\left(e^{\sigma^2} - 1\right)$

Maxwell Distribution
Range of X: $X > \theta$
Common use: The speed of a molecule in an ideal gas.

PDF: $f(x) = \sqrt{\dfrac{2}{\pi}} \dfrac{(x-\theta)^2}{\beta^3} \exp\left[-\dfrac{1}{2}\left(\dfrac{x-\theta}{\beta}\right)^2\right]$

Parameters: Scale $\beta > 0$, threshold θ
Mean: $\theta + \beta\sqrt{8}/\sqrt{\pi}$
Variance: $\beta^2(3 - 8/\pi)$

Noncentral Chi-square Distribution

Range of X: $X \geq 0$

Common use: Used to calculate the power of chi-square tests.

PDF: $f(x) = \sum_{j=0}^{\infty} \sum_{j=0}^{\infty} \frac{1}{j!} \left(\frac{c}{2} \right)^j e^{-c/2} \frac{x^{v/2+j-1} e^{-x/2}}{2^{v/2+j} \Gamma(v/2+j)}$

Parameters: Degrees of freedom $v > 0$, noncentrality $c \geq 0$

Mean: $v + c$

Variance: $2(v + 2c)$

Noncentral F Distribution

Range of X: $X \geq 0$

Common use: Used to calculate the power of F tests.

PDF: $f(x) = \sum_{j=0}^{\infty} \frac{1}{j!} \left(\frac{c}{2} \right)^j e^{-c/2}$

$$\times \frac{(v/w)^{v/2+j}}{B(v/2+j, w/2)} \left(1 + \frac{v}{w} x \right)^{-((v+w)/2+j)} x^{v/z+j-1}$$

Parameters: Numerator degrees of freedom $v > 0$, denominator degrees of freedom $w > 0$, noncentrality $c > 0$

Mean: $\dfrac{w(v+c)}{v(w-2)}$ if $w > 2$

Variance: $2 \left(\dfrac{w}{v} \right)^2 \dfrac{(v+c)^2 + (v+2c)(w-2)}{(w-2)^2 (w-4)}$ if $w > 4$

Noncentral t Distribution

Range of X: All real X

Common use: Used to calculate the power of t tests.

PDF: $f(x) = \sum_{j=0}^{\infty} \frac{1}{j!} \left(c\sqrt{2} \right)^j e^{-c^2/2} \frac{\Gamma[(v+j+1)/2]}{\Gamma(v/2)\Gamma(1/2)}$

$$\times \frac{x^j}{v^{(j+1)/2}} \left(1 + \frac{x^2}{v} \right)^{-(v+j+1)/2}$$

Parameters: Degrees of freedom $v > 0$, noncentrality $c \geq 0$

Mean: $(v/2)^{1/2} \dfrac{\Gamma\left[(v-1)/2\right]}{\Gamma(v/2)} c$

Variance: $\dfrac{v}{v-2}\left(1+c^2\right) - \left\{(v/2)^{1/2} \dfrac{\Gamma\left[(v-1)/2\right]}{\Gamma(v/2)} c\right\}^2$

Normal Distribution

Range of X: All real X

Common use: Widely used for measurement data, particularly when variability is due to many sources.

PDF: $f(x) = \dfrac{1}{\sqrt{2\pi}\sigma} e^{-\frac{(x-\mu)^2}{2\sigma^2}}$

Parameters: Mean μ, standard deviation $\sigma > 0$

Mean: μ

Variance: σ^2

Pareto Distribution

Range of X: $X \geq 1$

Common use: Model for many socioeconomic quantities with very long upper tails.

PDF: $f(x) = cx^{-c-1}$

Parameters: Shape $c > 0$

Mean: $\dfrac{c}{c-1}$ if $c > 1$

Variance: $\dfrac{c}{(c-2)(c-1)^2}$ if $c > 2$

Pareto Distribution (2-Parameter)

Range of X: $X \geq \theta$

Common use: Distribution of socioeconomic quantities with a lower bound.

PDF: $f(x) = c\theta^c x^{-c-1}$

Parameters: Shape $c > 0$, threshold $\theta > 0$

Mean: $\dfrac{\theta c}{c-1}$ if $c > 1$

Variance: $\dfrac{\theta^2 c}{(c-2)(c-1)^2}$ if $c > 2$

Rayleigh Distribution

Range of X: $X > \theta$

Common use: The distance between neighboring items in a pattern generated by a Poisson process.

PDF: $f(x) = \dfrac{2}{x-\theta}\left(\dfrac{x-\theta}{\beta}\right)^2 \exp\left[-\left(\dfrac{x-\theta}{\beta}\right)^2\right]$

Parameters: Scale $\beta > 0$, threshold θ

Mean: $\theta + \beta\sqrt{\pi}/2$

Variance: $\beta^2(1-\pi/4)$

Smallest Extreme Value (Gembel) Distribution

Range of X: All real X

Common use: Distribution of the smallest value in a sample from many distributions. Also used for negatively skewed measurement data.

PDF: $f(x) = \dfrac{1}{\beta}\exp\left\{\left(\dfrac{x-\gamma}{\beta}\right) - \exp\left(\dfrac{x-\gamma}{\beta}\right)\right\}$

Parameters: Mode γ, scale $\beta > 0$

Mean: $\gamma - 0.57722\beta$

Variance: $\dfrac{\beta^2\pi^2}{6}$

Student's t Distribution

Range of X: All real X

Common use: Reference distribution for the sample mean when sampling from a normal population with unknown variance.

PDF: $f(x) = \dfrac{\Gamma\left(\dfrac{v+1}{2}\right)\left[1+\dfrac{x^2}{v}\right]^{-(v+1)/2}}{\sqrt{\pi v}\,\Gamma\left(\dfrac{v}{2}\right)}$

Parameters: Degrees of freedom $v \geq 1$

Mean: 0

Variance: $\dfrac{v}{v-2}$ if $v > 2$

Triangular Distribution

Range of X: $a \leq X \leq b$

Common use: Often used as a rough model in the absence of data.

PDF:
$$f(x) = \frac{2(x-a)}{(b-a)(c-a)}, \quad x \leq c$$

$$f(x) = \frac{2(b-x)}{(b-a)(b-c)}, \quad x \geq c$$

Parameters: Lower limit a, mode $c \geq a$, upper limit $b \geq c$

Mean: $\dfrac{a+b+c}{3}$

Variance: $\dfrac{a^2 + b^2 + c^2 - ab - ac - bc}{18}$

U Distribution

Range of X: $b - a \leq X \leq b + a$

Common use: Used in metrology for the distribution of quantities that oscillate around a specific value.

PDF: $f(x) = \dfrac{1}{\pi a \sqrt{1 - \left(\dfrac{x-b}{a}\right)^2}}$

Parameters: Scale $a > 0$, mean b

Mean: b

Variance: $\dfrac{a^2}{2}$

Uniform Distribution

Range of X: $a \leq X \leq b$

Common use: Model for variable with equal probability everywhere over an interval.

PDF: $f(x) = \dfrac{1}{b-a}$

Parameters: Lower limit a, upper limit $b \geq a$

Mean: $\dfrac{a+b}{2}$

Variance: $\dfrac{(b-a)^2}{12}$

Weibull Distribution

Range of X: $X \geq 0$

Common use: Widely used in reliability analysis to model product lifetimes.

PDF: $f(x) = \dfrac{\alpha}{\beta^{\alpha}} x^{\alpha-1} e^{-(x/\beta)^{\alpha}}$

Parameters: Shape $\alpha > 0$, scale $\beta > 0$

Mean: $\dfrac{\beta}{\alpha} \Gamma\left(\dfrac{1}{\alpha}\right)$

Variance: $\dfrac{\beta^2}{\alpha}\left[2\Gamma\left(\dfrac{2}{\alpha}\right) - \dfrac{1}{\alpha}\Gamma\left(\dfrac{1}{\alpha}\right)^2\right]$

Weibull Distribution (3-Parameter)

Range of X: $X \geq \theta$

Common use: Used for product lifetimes with a lower bound.

PDF: $f(x) = \dfrac{\alpha}{\beta^{\alpha}} (x-\theta)^{\alpha-1} \exp\left[-(x-\theta)/\beta\right]^{\alpha}$

Parameters: Shape $\alpha > 0$, scale $\beta > 0$, threshold θ

Mean: $\theta + \dfrac{\beta}{\alpha} \Gamma\left(\dfrac{1}{\alpha}\right)$

Variance: $\dfrac{\beta^2}{\alpha}\left[2\Gamma\left(\dfrac{2}{\alpha}\right) - \dfrac{1}{\alpha}\Gamma\left(\dfrac{1}{\alpha}\right)^2\right]$

Appendix B: Guide to Capability Analysis Procedures in Statgraphics

This appendix indicates the procedures in Statgraphics Version 18 that may be used to calculate the various statistics described in this book. To reproduce each example, follow the step-by-step instructions given for that example. In some cases, you may need to use the *Graphics Options* dialog box to adjust the axis scaling to match that shown in the text. The sample data sets and other information may be found at www.statgraphics.com/process-capability-analysis-book.

Chapter 1: Introduction

Example 1.1 Medical Devices

Load sample data file *meddevices.sgd*. Select *Describe – Numeric Data – One Variable Analysis*. On the data input dialog box, enter "diameter" in the *Data* field. On the list of tables and graphs, select *Analysis Summary*, *Summary Statistics*, *Box-and-Whisker Plot*, and *Frequency Histogram*.

On the Pane Options dialog box for *Summary Statistics*, select *Average, Median, Standard Deviation, Coeff. of Variation, Minimum, Maximum, Range, Lower Quartile, Upper Quartile, Interquartile Range, Stnd. Skewness,* and *Stnd. Kurtosis.* On the Pane Options dialog box for *Frequency Histogram*, set *Number of Classes* to "40", *Lower Limit* to "1.9", and *Upper Limit* to "2.1".

Example 1.2 Airline Accidents

Load sample data file *accidents.sgd.* Select *Plot – Scatterplots – X-Y Plot.* On the data input dialog box, enter "Fatal Accidents/ (Flight Hours/100000)" in the *Y* field and "Year" in the X field. Once the graph is displayed, double-click to maximize it, push the *Smooth/Rotate* button on the analysis toolbar, set *Type* to "Robust Lowess", and set *Smoothing Fraction* to "50"%.

Chapter 2: Capability Analysis Based on Proportion of Nonconforming Items

Example 2.1 Estimating the Proportion of Nonconforming Items

Select *Describe – Numeric Data – Hypothesis Tests.* On the first dialog box, set the *Parameter* field to "Binomial Proportion". Set the *Sample Proportion* to "0.0" and the *Sample Size* field to "100". On the second dialog box, set the *Alternative Hypothesis* to "Not Equal" and *Alpha* to "5"% to create a two-sided confidence interval. To create an upper one-sided bound, select Analysis Options and set the *Alternative Hypothesis* to "Less Than".

To plot the likelihood function, select *Statlets – Statistical Modeling – Process Capability Analysis – Attributes.* Set *Parameter* to "proportion of nonconforming items", *Method* to "classical", *Number of nonconforming items* to "0", *Sample size* to "100", and *Confidence limits* to "Upper" at "95%".

Example 2.2 Sample Size Determination If No Defects Expected

Select *Statlets – Statistical Modeling – Process Capability Analysis – Attributes*. Set *Parameter* to "proportion of nonconforming items", *Method* to "classical", *Number of nonconforming items* to "0", and *Confidence limits* to "Upper" at "95%". Push the *Solve for n* button. Enter "0.001" in the *Target upper bound* field and select "Number of nonconformities (x)" for *Hold unchanged.*

Example 2.3 Using a Uniform Prior

Select *Statlets – Statistical Modeling – Process Capability Analysis – Attributes*. Set *Parameter* to "proportion of nonconforming items", *Method* to "Bayesian", *Number of nonconforming items* to "0", *Sample size* to "100", and set both *Beta prior parameters* to "1.0".

Example 2.4 Using an Informative Prior

Select *Statlets – Statistical Modeling – Process Capability Analysis – Attributes*. Set *Parameter* to "proportion of nonconforming items", *Method* to "Bayesian", *Number of nonconforming items* to "0", and *Sample size* to "100". Press the *Set prior* button. Set *Input* to "Two percentiles", the *Percentage* fields to "50" and "90", and the *Percentile* fields to "0.005" and "0.01".

Chapter 3: Capability Analysis Based on Rate of Nonconformities

Example 3.1 Estimating Aircraft Accident Rates

Select *Statlets – Statistical Modeling – Process Capability Analysis – Attributes*. Set *Parameter* to "mean rate of nonconformities", *Method* to "classical", *Number of nonconformities* to "3", *Sample size* to "88727934", and *Confidence limits* to "Upper". Push the Update button.

Example 3.2 Estimating Warranty Repair Rates

Select *Statlets – Statistical Modeling – Process Capability Analysis – Attributes*. Set *Parameter* to "mean rate of nonconformities", *Method* to "classical", *Number of nonconformities* to "65", *Sample size* to "1000", and *Confidence limits* to "Upper". Push the Update button.

Example 3.3 Sample Size Determination

Select *Statlets – Statistical Modeling – Process Capability Analysis – Attributes*. Set *Parameter* to "mean rate of nonconformities", *Method* to "classical", *Number of nonconformities* to "65", *Sample size* to "1000", and *Confidence limits* to "Upper". Press the *Solve for n* button. Set *Target upper bound* to "0.07" and *Hold unchanged* to "Point estimate (x/n)".

Example 3.4 Bayesian Estimation of Fatal Accident Rate

Select *Statlets – Statistical Modeling – Process Capability Analysis – Attributes*. Set *Parameter* to "mean rate of nonconformities", *Method* to "Bayesian", *Number of nonconformities* to "3", *Sample size* to "88727934", and *Bayesian limits* to "Upper". Press the *Set prior* button. Set *Input* to "Two percentiles", the *Percentage* fields to "50" and "90", and the *Percentile* fields to "3e-8" and "5e-8".

Chapter 4: Capability Analysis of Normally Distributed Data

Example 4.1 Fitting a Normal Distribution

Load sample data file *meddevices.sgd*. Select *SPC – Capability Analysis – Variables – Individuals*. On the data input dialog box, enter "diameter" into the *Data* field. Enter "1.9" in the *LSL* field, "2.0" in the *Nominal* field, and "2.1" in the *USL* field. On the Analysis Options dialog box, set *Distribution* to "Normal". Maximize the *Capability Plot* by double-clicking on it. Select Pane Options. Set the *Number of Classes* to "40", the *Lower Limit* to "1.9", the *Upper Limit* to "2.1", and check *Hold*.

To calculate confidence limits, select *Describe – Numeric Data – One Variable Analysis*. On the data input dialog box, enter "diameter" into the *Data* field. On the list of tables and graphs, select *Confidence Intervals*.

Example 4.2 Analysis of Subgroup Data

Load sample data file *meddevices.sgd*. Select *SPC – Capability Analysis – Variables – Grouped Data*. On the data input dialog box, enter "diameter" into the *Data* field and "5" in the *Date/ Time/Labels or Size* field. Enter "1.9" in the *LSL* field, "2.0" in the *Nominal* field, and "2.1" in the *USL* field. On the Analysis Options dialog box, set *Distribution* to "Normal". On the list of tables and graphs, select *Tolerance Chart*. Maximize the tolerance chart by double-clicking on it. To jitter the data, push the *Jitter* button on the analysis toolbar and add a small amount of *Horizontal* jitter.

To create the box-and-whisker plot, select *Plot – Exploratory Plots – Box-and-Whisker Plots – Multiple Samples*. On the data input dialog box, enter "diameter" into the *Data* field and "REP(COUNT(1,20,1),5)" in the *Level codes* field. Use Pane Options to set *Direction* to "Vertical" and uncheck *Outlier Symbols*. Use Graphics Options to rescale the Y-axis so that it ranges from "1.85" to "2.15" by "0.05" and check the *Hold Sealing Constant* box. Use the *Add Object* button on the analysis toolbar to add *Horizontal lines* at "1.9", "2.0", and "2.1".

Example 4.3 Estimating Short-Term and Long-Term Variability from Subgroups

Load sample data file *meddevices.sgd*. Select *Edit – Preferences*. On the *Capability* tab, set the *Short-term sigma – grouped data* field to the desired setting. Check *Apply bias correction for s* if desired. Then select *SPC – Capability Analysis – Variables – Grouped Data*. On the data input dialog box, enter "diameter" into the *Data* field and "5" in the *Date/ Time/Labels or Size* field. Enter "1.9" in the *LSL* field, "2.0" in the *Nominal* field, and "2.1" in the *USL* field. Select *Capability Indices* from the list of tables and graphs.

Example 4.4 Estimating Short-Term Variability from Individuals Data

Load sample data file *meddevices.sgd*. Select *Edit – Preferences*. On the *Capability* tab, set the *Short-term sigma – individuals* field to the desired setting. Check *Apply bias correction for s* if desired. Then select *SPC – Capability Analysis – Variables – Individuals*. On the data input dialog box, enter "diameter" into the *Data* field. Enter "1.9" in the *LSL* field, "2.0" in the *Nominal* field, and "2.1" in the *USL* field. Select *Capability Indices* from the list of tables and graphs.

To create the MR(2) chart, select *SPC – Control Charts – Basic Variables Charts – Individuals*. On the data input dialog box, enter "diameter" in the *Observations* field. Select *MR(2) Chart* from the list of tables and graphs. Use Pane Options to set *Decimal Places for Limits* to "4".

Example 4.5 Capability Analysis of Medical Device Diameters

Load sample data file *meddevices.sgd*. Select *SPC – Capability Analysis – Variables – Individuals*. On the data input dialog box, enter "diameter" into the *Data* field. Enter "1.9" in the *LSL* field, "2.0" in the *Nominal* field, and "2.1" in the *USL* field. Select *Analysis Summary, Capability Indices,* and *Capability Plot* from the list of tables and graphs. On the Pane Options dialog box for *Capability Indices*, select the desired capability indices and set *Confidence Limits* to "Lower Confidence Bounds".

Example 4.6 Confidence Limits for One-Sided Specifications

Load sample data file *meddevices.sgd*. Select *SPC – Capability Analysis – Variables – Individuals*. On the data input dialog box, enter "diameter" into the *Data* field. Enter "1.9" in the *LSL* field but leave the *USL* field blank. Select *Analysis Summary*, *Capability Indices,* and *Capability Plot* from the list of tables and graphs. On the Pane Options dialog box for *Capability Indices*, select the desired capability indices and set *Confidence Limits* to "Lower Confidence Bounds".

Example 4.7 Confidence Limits for Two-Sided Specifications

Load sample data file *meddevices.sgd*. Select *SPC – Capability Analysis – Variables – Individuals*. On the data input dialog box, enter "diameter" into the *Data* field. Enter "1.9" in the *LSL* field, "2.0" in the *Nominal* field, and "2.1" in the *USL* field. Select *Analysis Summary*, *Capability Indices,* and *Capability Plot* from the list of tables and graphs. On the Pane Options dialog box for *Capability Indices*, select all of the indices to *Display* and set *Confidence Limits* to "Lower Confidence Bounds". Check *Include Bootstrap* and set the *Number of Subsamples* to "10000". Because at the random element involved in bootstrapping, the results, may not match exactly those shown in the text.

Chapter 5: Capability Analysis of Nonnormal Data

Example 5.1 Tests of Normality

Load sample data file *meddevices.sgd*. Select *Statlets – Statistical Modeling – Power Transformations*. On the data input dialog box, enter "diameter" into the *Data* field.

Example 5.2 Power Transformations

Repeat Example 5.1. To optimize the power only, push the *Optimize* button. To optimize both the power and the addend, check *Optimize addend* and then push the *Optimize* button.

Example 5.3 Calculating Process Capability for Transformed Data

Load sample data file *meddevices.sgd*. Select *Statlets – Statistical Modeling – Process Capability Analysis – Variables*. On the data input dialog box, enter "diameter" into the *Data* field. Enter "1.9" in the *LSL* field, "2.0" in the *Nominal* field, and "2.1" in the *USL* field. Once the Statlet window opens, set the *Addend* field to "-1.78688" and *Power* to "-3.15".

To calculate confidence bounds, select *SPC – Capability Analysis – Variables – Individuals*. On the data input dialog box, enter "diameter" into the *Data* field. Enter "1.9" in the *LSL* field, "2.0" in the *Nominal* field, and "2.1" in the *USL* field. On the Analysis Options dialog box, set *Distribution* to "Normal" and *Data Transformation* to "Power". Set the *Power* to "-3.15" and *Lower Threshold* to "1.78688". On the list of tables and graphs, select *Capability Indices*. On the *Capability Indices* Pane Options dialog box, select all of the capability indices and set *Confidence Limits* to "Lower Confidence Bounds".

Example 5.4 Fitting an Alternative Distribution

Load sample data file *meddevices.sgd*. Select *Describe – Distribution Fitting – Fitting Uncensored Data*. On the data input dialog box, enter "diameter" into the *Data* field. On the list of tables and graphs, select *Comparison of Alternative Distributions*. Maximize the *Comparison of Alternative Distributions* pane. On the Pane Options dialog box, select the distributions listed in Table 5.4. Also, press the *Tests* button. Under *Include*, select "Likelihood", "Kolmogorov-Smirnov D", and "Anderson-Darling A^2". Under *Sort by*, select "Anderson-Darling A^2".

To plot the best-fitting distributions, display the Analysis Options dialog box and check "Generalized Logistic", "Largest Extreme Value", and "Loglogistic (3-parameter)".

To perform the capability analysis, select *SPC – Capability Analysis – Variables – Individuals*. On the data input dialog box, enter "diameter" into the *Data* field. Enter "1.9" in the *LSL* field, "2.0" in the *Nominal* field, and "2.1" in the *USL* field. On the Analysis Options dialog box, set *Distribution* to "Largest Extreme Value".

Example 5.5 Testing Goodness-of-Fit of Nonnormal Distribution

Load sample data file *meddevices.sgd*. Select *Describe – Distribution Fitting – Fitting Uncensored Data*. On the data input dialog box, enter "diameter" into the *Data* field. On the Analysis Options dialog box, set *Distribution* to "Largest Extreme Value". On the list of tables and graphs, select *Goodness-of-Fit Tests*. On the Pane Options dialog box for *Goodness-of-Fit Tests, check* "Modified Kolmogorov-Smirnov D" and "Anderson-Darling A^2". Check "Calculate distribution-specific P-values".

Example 5.6 Calculating Capability Indices for Nonnormal Distribution

Load sample data file *meddevices.sgd*. Select *SPC – Capability Analysis – Variables – Individuals*. On the data input dialog box, enter "diameter" into the *Data* field. Enter "1.9" in the *LSL* field, "2.0" in the *Nominal* field, and "2.1" in the *USL* field. On the Analysis Options dialog box, set *Distribution* to "Largest extreme value". On the list of tables and graphs, select *Analysis Summary* and *Capability Indices*. On the Pane Options dialog box for *Capability Indices*, select all of the capability indices, set *Confidence Limits* to "Lower Confidence Bounds", and set the *Number of Subsamples* to "50000".

Example 5.7 Transformation Using Johnson Curves

Load sample data file *meddevices.sgd*. Select *Statlets – Statistical Modeling – Process Capability Analysis – Variables*. On the data input dialog box, enter "diameter" into the *Data* field. Enter "1.9" in the *LSL* field, "2.0" in the *Nominal* field, and "2.1" in the *USL* field. On the Statlets toolbar, set *Distribution* to "Johnson".

To calculate the capability indices, select *SPC – Capability Analysis – Variables – Individuals*. On the data input dialog box, enter "diameter" into the *Data* field. Enter "1.9" in the *LSL* field, "2.0" in the *Nominal* field, and "2.1" in the *USL* field. On the Analysis Options dialog box, set *Distribution* to "Johnson SB, SL, SU". On the list of tables and graphs, select *Analysis Summary* and *Capability Indices*. On the Pane Options dialog box for *Capability Indices*, select all of the capability indices, set *Confidence Limits* to "Lower Confidence Bounds", and set the *Number of Subsamples* to "50000".

Chapter 6: Statistical Tolerance Limits

Example 6.1 Analysis of Medical Device Diameters

Load sample data file *meddevices.sgd*. Select *Statlets – Statistical Modeling – Process Capability Analysis - Variables*. On the data input dialog box, enter "diameter" into the *Data* field. Enter "1.9" in the *LSL* field, "2.0" in the *Nominal* field, and "2.1" in the *USL* field. On the Statlet toolbar, select *Tolerance limits* and enter "95% for 99%". To create a one-sided limit, change "Two-sided" to "Upper".

Example 6.2 Use of Power Transformations

Repeat instructions for Example 6.1. Then enter "−1.78688" in the *Addend* field and set *Power* to −3.15.

Example 6.3 Tolerance Limits Based on Largest Extreme Value Distribution

Repeat instructions for Example 6.1. Then set *Distribution* field to "Largest extreme value".

Example 6.4 Nonparametric Tolerance Limits

Load sample data file *meddevices.sgd*. Select *Describe − Numeric Data − Statistical Tolerance Limits − From Observations*. On the data input dialog box, enter "diameter" into the *Data* field. Enter "1.9" in the *LSL* field and "2.1" in the *USL* field. On the Analysis Options dialog box, set *Distribution* to "Nonparametric (specified confidence)" with *Interval Depth* set to "1". Set *Confidence Level* to "95"%.

   To control the population proportion, use Analysis Options to change *Distribution* to "Nonparametric (specified proportion)" with *Interval Depth* set to "1". Set *Population Proportion* to "99"%.

Chapter 7: Multivariate Capability Analysis

Example 7.1 Bivariate Data Visualization

Load sample data file *devices.sgd*. To create the scatterplot, select *Plot − Scatterplots − X-Y Plot*. On the data input dialog box, enter "strength" for *Y* and "diameter" for *X*. Use *Graphics Options* to rescale the X-axis to range from 1.85 to 2.15 by 0.05 and the Y-axis to range from 180 to 280 by 10. Use the *Add Object* button to add line segments from

(1.9,200) to (1.9,280), from (2.1,200) to (2.1,280), and from (1.9,200) to (2.1,200).

To create the bivariate histogram, select *Statlets – Data Exploration – Bivariate Density*. On the data input dialog box, enter "diameter" for *Sample 1* and "strength" for *Sample 2*. To switch to a nonparametric density estimate, select *Nonparametric density estimate with width* set equal to "30%" and then set the *Resolution* field to "201".

Example 7.2 Fitting a Multivariate Normal Distribution

Load sample data file *devices.sgd*. Select *Statlets – Data Exploration – Bivariate Density*. On the data input dialog box, enter "diameter" for *Sample 1* and "strength" for *Sample 2*. On the Statlet toolbar, select "Normal distribution" and set the *Resolution* field to "201".

Example 7.3 Tests for Multivariate Normality

Load sample data file *devices.sgd*. Select *Describe – Multivariate Methods – Multivariate Normality Test*. On the data input dialog box, enter "diameter" and "strength" in the *Data* field.

Example 7.4 Multivariate Capability Indices

Load sample data file *devices.sgd*. Select *SPC – Capability Analysis – Variables – Multivariate Capability Analysis*. On the data input dialog box, enter "diameter" and "strength" in the *Data* field. Enter "USL" in the *Upper Specification Limits* field, "Nominal" in the *Nominal Values* field, and "LSL" in the *Lower Specification Limits* field. Select *Analysis Summary, Capability Indices, Capability Plot*, and *Capability Ellipse* from the list of tables and graphs. On the Pane

Options dialog box for the *Capability Indices,* set *Bootstrap Confidence Limits* to "Lower confidence bounds" and *Number of Subsamples* to "5000".

Example 7.5 Multivariate Normal Tolerance Region

Load sample data file *devices.sgd.* Select *Describe – Numeric Data – Statistical Tolerance Limits – Multivariate Tolerance Limits.* On the data input dialog box, enter "diameter" and "strength" in the *Data* field. On the Analysis Options dialog box, set *Confidence Level* to "95"% and *Population Proportion* to "99.9"%. Move "strength" to the *Lower bound only* field. Use *Graphics Options* to rescale the X-axis to range from 1.85 to 2.15 by 0.05 and the Y-axis to range from 180 to 300 by 20. Use the *Add Object* button to add line segments from (1.9,200) to (1.9,300), from (1.9,200) to (2.1,200), and from (2.1,200) to (2.1,300). Add the text string "1.9" at position (1.9,300) and the text string "2.1" at position (2.1,300) after setting *Properties – Reference Position* to "Bottom center".

Example 7.6 Analyzing Multivariate Lognormal Data

Load sample data file *rmlognormal.sgd.* Select *Describe – Multivariate Methods – Multivariate Normality Test.* On the data input dialog box, enter "X1" and "X2" in the *Data* field. On the Analysis Options dialog box, set *Transformation* to "Multivariate power transformation".

To create the bivariate density estimate for the transformed data, select *Statlets – Data Exploration – Bivariate Density.* On the data input dialog box, enter "X1^-0.08127" for *Sample 1* and "X2^0.0733825" for *Sample 2.* Set the *Resolution* field to "201" and then select *Nonparametric density estimate with width* set equal to "50%".

Chapter 8: Sample Size Determination

Example 8.1 Sample Size Determination for Proportion Nonconforming

Select *Tools – Sample Size Determination – One Sample*. On the first dialog box, select *Binomial Proportion* and enter "0.001" in the *Hypothesized Proportion* field. On the second dialog box, set the *Control* field to "Absolute Error" and enter "0.0002". Press OK.

To specify beta risk, return to the Analysis Options dialog box. Change the *Control* field to "Power", set the power to "90%", set the *Difference to Detect* field to "0.001", and set the *Alternative Hypothesis* to "Greater Than".

Example 8.2 Sample Size Determination for Rate of Nonconformities

Select *Tools – Sample Size Determination – One Sample*. On the first dialog box, select *Poisson Rate* and enter "3.0" in the *Hypothesized Rate* field. On the second dialog box, set the *Control* field to "Relative Error" and enter "20.0"%. Set the *Alternative Hypothesis* to "Greater Than" and press OK.

Example 8.3 Sample Size Determination for C_p

Select *Tools – Sample Size Determination – Capability Indices*. On the input dialog box, select C_p. Set the *Relative error* field to "10"% and the *Confidence Level* to "95"%.

Example 8.4 Sample Size Determination for C_{pk}

Select *Tools – Sample Size Determination – Capability Indices*. On the input dialog box, select C_{pk}. Set the *Estimated index* to "1.33", the *Relative error* field to "10"%, and the *Confidence Level* to "95"%.

Example 8.5 Sample Size Determination for Statistical Tolerance Limits

Select *Tools – Sample Size Determination – Statistical Tolerance Limits*. On the input dialog box, set *Distribution* to "Largest extreme value", the *Mode* parameter to "2.0", the *Scale* parameter to "0.015", the *Type of Limits* to "Two-sided", the *Confidence Level* to "95"%, the *Population Proportion* to "99"%, the *Lower Spec. Limit* to "1.9", and the *Upper Spec. Limit* to "2.1". To control the Monte Carlo simulation, set the *Inclusion Percentage* to "90"%, the *Number of Trials* to "50000", and the *Maximum n* to "1000". Be patient waiting for the results.

Chapter 9: Control Charts for Process Capability

Example 9.1 Control Chart for Proportion of Nonconforming Items

Load sample data file *capcontrol.sgd*. Select *SPC – Control Charts – Capability Control Charts – Attributes*. On the data input dialog box, enter "Number nonconforming/sample size" in the *Statistic* field and "Sample size" in the *Sample size or sizes* field. On the Analysis Options dialog box, set *Parameter* to "Proportion", *Target mean* to "0.01", and *Limits* to "Upper only". On the list of tables and graphs, select *Analysis Summary, Capability Chart,* and *OC Curve*.

Example 9.2 Control Chart for Rate of Nonconformities

Load sample data file *capcontrol.sgd*. Select *SPC – Control Charts – Capability Control Charts – Attributes*. On the data input dialog box, enter "Accidents/Departures" in the *Statistic* field, "Departures" in the *Sample size or sizes* field and "Year" in the *Date/Time/Labels* field. On the Analysis Options dialog box, set *Parameter* to "Rate", *Target mean* to "3.5", and *Limits* to "Upper and Lower". On the list of tables and graphs, select *Analysis Summary* and *Capability Chart.* On the Pane Options dialog box for *Capability Chart,* check *Plot outer warning limits, Plot inner warning limits,* and *Mark runs rules violations,* and set *Decimal places for limits* to "2".

Example 9.3 Control Chart for C$_p$

Load sample data file *capcontrol.sgd.* Select *SPC – Control Charts – Capability Control Charts – Variables.* On the data input dialog box, enter "Cp" in the *Capability index* field and "30" in the *Sample size or sizes* field. On the Analysis Options dialog box, set *Parameter* to "Cp (short-term)", *Target index* to "2.0" and *Limits* to "Upper and lower". On the list of tables and graphs, select *Analysis Summary* and *Capability Chart.*

Example 9.4 Control Chart for C$_{pk}$

Load sample data file *capcontrol.sgd.* Select *SPC – Control Charts – Capability Control Charts – Variables.* On the data input dialog box, enter "Cpk" in the *Capability index* field and "30" in the *Sample size or sizes* field. On the Analysis Options dialog box, set *Parameter* to "Cpk (short-term)", *Target index* to "1.5", and *Limits* to "Upper and lower". On the list of tables and graphs, select *Analysis Summary, Capability Chart,* and *OC Curve.*

Example 9.5 Sample Size Determination for C$_{pk}$ Control Chart

Select *Statlets – Sampling – Capability Control Chart Design.* Set the *Parameter to be estimated* field to "Cpk (short-term)". Set the *Base sample size on* field to "Power". Set the *Centerline* field to "1.5" and the *Alternative value field* to "1.0". Set the *Alpha risk* field to "0.5%" and the *Power* field to "90%". Set the *Chart type* field to "One-sided".

Example 9.6 Acceptance Control Charts

Load sample data file *diameterxbars.sgd*. To create the X-bar chart, select *SPC – Control Charts – Basic Variables Charts – Xbar and S.* Select *Subgroup statistics* on the data input dialog box and enter "Xbar", "s", and "30" in the input fields. On the Analysis Options dialog box, set the *Type of Study* to "Control to Standard", set *Mean* to "2.0", and set *Std. Dev.* to "0.01".

To create the acceptance control chart, select *SPC – Control Charts – Special Purpose Control Charts – Acceptance Chart.* Select *Subgroup statistics* on the data input dialog box and enter "Xbar", "s", and "30" in the input fields. Set *LSL* to "1.9" and *USL* to "2.1". On the Analysis Options dialog box, set the *Type of Study* to "Control to Standard" and set the *Mean and Std. Dev.* equal to "2.0" and "0.01", respectively. Set *Specify* to "Sigma Multiple". Set *Fraction Nonconforming* to "0.0001" and *Sigma multiple* to "3.0".

To change to the beta risk method, return to *Analysis Options* and set *Specify* to "Beta Risk", *Fraction Nonconforming* to "0.001", and *Beta risk* to "0.05".

Note: When using the Acceptance Control Charts with subgroup statistics, the data input dialog box will request either "Standard deviations" or "Ranges", depending on how the system preference *Preferred dispersion chart* is set.

Example 9.7 Sample Size Determination for Acceptance Control Charts

Press the *Evaluator* button on the main toolbar. Enter the following expression:

$$\Big(\big(\text{invnormal}\,(1 - .0027, 0, 1) + \text{invnormal}\,(1 - .05, 0, 1)\big) /$$

$$\big(\text{invnormal}\,(1 - .0001, 0, 1) - \text{invnormal}\,(1 - .001, 0, 1)\big)\Big) \wedge 2$$

Default Preferences

Statgraphics Version 18 also maintains a set of user preferences that affect some of the calculations performed in this book. To reproduce the results, the reader should select *Edit – Preferences* from the main menu and set the preferences as indicated below.

General tab
  Confidence Level: 95%
  Significant Digits: 6
Dist. Fit tab
  Tests for Normality: Shapiro-Wilk
  General Goodness-of-Fit Tests: Anderson-Darling A^2
  Calculate distribution-specific P-values: checked.

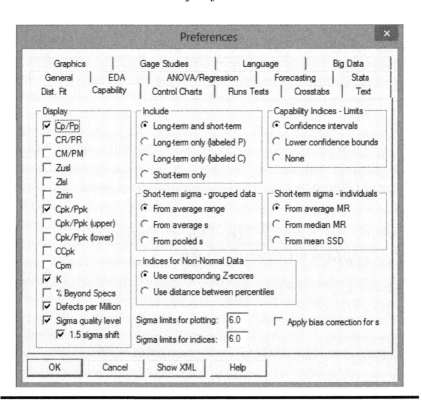

Figure B.1   Default preferences for calculation of capability indices.

Control Charts tab
   Preferred Dispersion Chart: Sigma
   Sigma multiple: 3.0
Capability tab
   The options selected on this tab affect which capability
      indices are displayed by default and what estimates of
      sigma are used to calculate them. The settings selected
      in Figure B.1 are consistent with common practice.

Index

Q

Printed in the United States
by Baker & Taylor Publisher Services